THE BIOCHEMICAL BASIS
OF SPORTS PERFORMANCE

The Biochemical Basis of Sports Performance

SECOND EDITION

Ron Maughan & Michael Gleeson

OXFORD
UNIVERSITY PRESS

OXFORD

UNIVERSITY PRESS

Great Clarendon Street, Oxford ox2 6DP

Oxford University Press is a department of the University of Oxford.
It furthers the University's objective of excellence in research, scholarship,
and education by publishing worldwide in

Oxford New York

Auckland Cape Town Dar es Salaam Hong Kong Karachi
Kuala Lumpur Madrid Melbourne Mexico City Nairobi
New Delhi Shanghai Taipei Toronto

With offices in

Argentina Austria Brazil Chile Czech Republic France Greece
Guatemala Hungary Italy Japan Poland Portugal Singapore
South Korea Switzerland Thailand Turkey Ukraine Vietnam

Oxford is a registered trade mark of Oxford University Press
in the UK and in certain other countries

Published in the United States
by Oxford University Press Inc., New York

First edition published 2004

Second edition published 2010

British Library Cataloguing in Publication Data

Data available

Library of Congress Cataloging in Publication Data

Data available

Typeset by MPS Limited, A Macmillan Company
Printed in Great Britain on acid-free paper by
Ashford Colour Press Ltd, Gosport, Hants

ISBN 978–0–19–920828–9

5 7 9 10 8 6 4

Preface to the Second Edition

It seems only a short time since the first edition of this book appeared. In reality, it is six years. In that time, there have been many major advances in the field of exercise biochemistry, and there have also been a few retreats. Many of these have come from the application of new techniques to the study of old problems, and from the cross-fertilization of ideas from different areas of specialization. Molecular biology continues to develop rapidly and to provide insights into the mechanisms operating at the sub-cellular level that underpin many of the responses and adaptations to different types of exercise stress. The cytokines, for example, are now recognized as having a wide range of roles in intracellular signalling pathways. Likewise, free radicals, formerly thought of as being exclusively harmful to cells, are seen to play important roles in a range of cellular processes. The rapid developments in gene therapy, with their implications for abuse in sport, offer the prospect of tissue engineering in ways that did not seem possible even a few years ago.

In revising the text, we have tried to take account of the new information that has emerged, and some substantial new sections have been added. At the same time though, we have tried not to lose sight of our original aim, which was 'to introduce students of sports science or exercise physiology to the biochemical processes that underpin exercise performance and the adaptations that occur with training.' As pointed out in the preface to the first edition, we have tried to maintain the focus on physiological chemistry—the processes that occur in the living body under physiological conditions rather than those that can occur in the test tube.

We have added many new references to the further reading lists at the end of each chapter to illustrate recent advances in knowledge. We have, however, also retained, and even added, many older references. We have done this as we believe that students should not lose sight of the development of the subject: many of these older references highlight major developments as they occurred and they are still relevant today.

This second edition has benefited from the many comments made by colleagues and former students. The fact that these have been generally positive has encouraged us in the preparation of a new edition.

Ron Maughan
Mike Gleeson
Loughborough University, 2010

Preface to the First Edition

Some understanding of the biochemistry of exercise is fundamental to any study of the factors that contribute to sports performance. It is the physical, chemical and biochemical properties of cells and tissues that determine the physiological responses to exercise, and yet the teaching of exercise biochemistry is poorly developed compared with exercise physiology. Where the subject is taught at all, the student often finds the approach somewhat daunting, with its focus on thermodynamics, chemical structures and metabolic pathways. Many students find the subject difficult, when it should not be so.

The aim of this book is to introduce the student of sports science or exercise physiology to the biochemical processes that underpin exercise performance and the adaptations that occur with training. The focus is on skeletal muscle metabolism and the provision of energy for working muscles. This is strictly physiological chemistry—the study of the biochemical or metabolic processes that occur in the whole organism—as opposed to those that might occur in the test-tube under non-physiological conditions. Although the factors that cause fatigue during exercise are not well understood, it seems likely that they have a biochemical rather than a physiological basis. Students of sport and exercise science do not require a deep understanding of biochemistry, but do need to be familiar with the main concepts.

We have tried in this book to introduce the principles of exercise biochemistry in a context that is immediately relevant to the student of sports science. This has meant abandoning the traditional approach of working through the main classes of biomolecules and the major metabolic pathways. Instead, we have tackled the subject by considering the biochemical processes involved in energy provision for different sports events and the way in which limitations in the energy supply can cause fatigue and thus limit performance. Recovery from exercise is important for athletes who train and compete with only a limited rest period, and the biochemical processes that influence recovery and restoration of performance capacity are also addressed in this book.

The subject matter lends itself well to this approach. The weightlifter is concerned only with force production, but the sprinter needs to sustain high-power outputs, albeit only for short periods of time. The endurance athlete must produce only a moderate power output, but this rate of work must be sustained for long periods. The biochemical processes that fuel the different activities that contribute to sport are the focus of this book, together with the changes that occur with training and the role of diet in providing the necessary fuels. Sporting talent is a rare gift inherited by the elite athlete from his or her parents, and a brief description of the basis of heredity is included.

The authors both teach undergraduate courses in exercise biochemistry, physiology and nutrition and bring their experience of research and teaching to this book in the hope that it will stimulate the interested student to pursue this fascinating subject.

Ron Maughan
Mike Gleeson

Contents

3   The sprinter

4 Middle-distance events

5 The endurance athlete

8   Adaptations to training

Appendix 1 Key concepts in physical, organic, and biological chemistry

Appendix 2 Glossary of abbreviations and biochemical terminology

Appendix 3 Units commonly used in biochemistry and physiology

Introduction:
The biochemical basis of exercise and sport

Learning objectives

After studying this chapter, you should be able to . . .

1. appreciate why knowledge of biochemistry is important to our understanding of the factors that determine success in sport;

2. describe some of the factors that have contributed to the improvement of world records in sports events over the last century;

3. understand the limitations of extrapolating from the changes in previous world records to predict the pace of future improvements;

4. describe some of the factors that limit exercise performance and determine success in sport;

5. describe the concept of failure of ATP supply as a biochemical cause of fatigue.

Introduction

The innate level of ability and the ability to respond to training with an improvement in performance are the keys to success in sport

All sports activities involve muscular activity, and for each one of us there is an upper limit to our ability to perform any task involving muscular effort. When we try to do more, fatigue intervenes: when this happens, we have to slow down or stop, and skill and coordination deteriorate. This applies to all activities involving strength, speed, stamina, and skill. It is these differences in physical capacity between individuals that form the basis of sporting contests as each competitor strives to reach those limits. The limitations to exercise performance are many and varied, but no matter what our event or our level of performance, we can improve with appropriate training. These two factors, the innate level of ability and the ability to respond to training with an improvement in performance, are the keys to success in sport.

The genetic material dictates the formation of proteins and these in turn control metabolism

Both of these factors are essentially determined by our body's biochemical make-up. Our genetic material, which is made up of deoxyribonucleic acid (DNA), dictates the formation of proteins and these in turn control the metabolism of all the other chemical elements and compounds that make up our cells and tissues. The human body is nothing more than a collection of biochemicals that interact with each other to provide both structure and function. The structure is important: it determines our height, weight, external appearance, and all our other physical characteristics. Function is even more important as this determines strength, speed, stamina, and skill. The nature of the mind, of thought processes or even of simpler human attributes such as memory, is not well understood, but all of these phenomena must also have a biochemical basis. Physiology and medicine are no more than the physical manifestations of the underlying biochemistry. Some would argue that psychology falls into the same category.

Understanding the body's responses to exercise is important for the athlete who seeks to perform well

Sport takes many different forms, and offers as many opportunities for those who simply want to participate in a pleasurable recreational or social activity as it does for those who want to compete at the highest level. The recreational participant can benefit from long-term improvements in physical and mental well-being and in functional capacity as well as the immediate pleasure of participation. Understanding the body's responses to exercise is important for the athlete who seeks to perform well, but it is also a crucial element in the development of successful physical activity programmes that can tackle the growing prevalence of the diseases that accompany a sedentary lifestyle.

An understanding of biochemistry—the chemistry of life—is fundamental to all of these aspects and it will become even more so as science progresses from simple descriptions of phenomena based on observation to a detailed understanding of the mechanisms that control bodily function.

Historical perspective

In team sports, there are few opportunities to set records, and the standards by which participants are judged are relative rather than absolute. The aim is simply to beat the opposition. Even in these sports, however, meticulous records of team and individual performances are kept, and the exceptional ability of outstanding individual players is recognized. Baseball players such as Joe DiMaggio, cricketers such as Don Bradman, and Pele from the world of soccer are remembered for the statistics they generated, but much more for their individual brilliance. To the younger generation this is seen only from film or television recordings, but memories of earlier days, before film records were kept, still persist. The great cricketer WG Grace, who was at his peak in the 1870s and 1880s, was filmed only in his declining years, and the few photographs available can convey little of his skills. We know, however, that Walter George set a world record for the mile in 1886 and that his time was recorded as 4 min 12.75 s. That may sound slow compared with today's performances, but it remains a creditable performance, given the state of the tracks on which he ran, the shoes he wore, and the limited training that he undertook.

Evolution of records

In some sports, the rules or the equipment have changed, making historical comparisons inappropriate

In sports where performance can be measured, it is possible to record the evolution of records, and the world record is a measure of the limit of what was achieved at any particular time in the history of sport. In some sports, the rules have changed or the equipment has changed, making historical comparisons inappropriate. The length and weight of the javelin, for example, were changed when the distances achieved by the top throwers became so great that they could not be contained within the confines of an athletics stadium. Likewise, the introduction of aluminium and then fibreglass poles in the pole vault has made historical comparisons inappropriate. Most events, however, have not undergone fundamental changes, although it could be argued that the introduction of synthetic running tracks and improved shoe design have altered even the simple activity of running from A to B as fast as possible.

> In the 100-m sprint, the men's record has always been better than the women's record

In an event such as the 100-m sprint on the track, successful performers must react rapidly to the starter's gun and generate high power to accelerate from the blocks. They must then maintain peak speed for as long as possible and slow down as little as possible in the later stages of the race. Improvements in performance may come from faster reaction times, better ability to generate power in the active muscles, improved resistance to fatigue, or some combination of these. Figure 1.1 shows how performances in the 100-m sprint have improved for men and women since records were first officially recognized in 1912. A number of factors are apparent from these graphs:

- there has been a steady improvement in performances for as long as records have been kept;
- there is no obvious indication that the rate of improvement has slowed—progress has been more or less linear at a rate of about 0.1 s per decade;
- the men's record has always been better than the women's record and even though the gap has narrowed somewhat, there is still a large (about 10%) difference.

Figure 1.1

Evolution of world record times for the 100-m sprint for men and women since official record keeping began.

> There must be a limit to performance and at some point in the future there will be no further improvement

Although much can be learned from looking at a simple graph such as Figure 1.1, it is easy to over-interpret or misinterpret the information and to reach inappropriate conclusions. It is clear that there must be a limit to performance and at some point in the future there will be no further improvement. If we consider only the data for men, we can apply a linear regression analysis, which shows that there is no sign of the rate of progress falling off; the correlation coefficient is 0.948. This predicts that the world record will be 9.62 s in 2020 and 8.99 s in 2100 (Figure 1.2a). Clearly, however, the equation is not quite right as the World Record already exceeds that time—Usain Bolt having run a time of 9.58 s in 2009. That is the trouble with this type of analysis—progression is not linear, with periods of rapid improvement and other times when performances seem to stagnate. The women's record has not improved since 1988, a period of more than 20 years. It is clear that this time will be broken at some stage, but there is no way of knowing when this will happen. Another problem with the linear analysis model is the assumption that performances will continue to improve. A time of 8.99 s for the men's 100 m seems possible, but extending this extrapolation further suggests that at some stage in the future (in fact in the year 3276) the world record will be 0.00 s. The year after it will be negative—i.e. the runner who wants to break this record must finish before he has started. This is clearly absurd and will not happen, illustrating some of the difficulties with projections based on linear analyses. However, if we analyse these data differently and apply a non-linear equation to fit the best line, this fits just as well and indicates a gradual slowing of the rate of improvement. This analysis of the same information predicts that the ultimate limit will lie somewhere around 9.2 s, (Figure 1.2b) but it fails because it does not allow for an athlete like Usain Bolt, who was responsible for lowering the record by 0.16 s in less than 2 years. Only time will tell where the limit lies, and we cannot predict what impact new developments in track surfaces, shoes, and other factors may have on performance. Most predictions have been proved wrong by subsequent developments, but it seems safe to say that no man or woman will ever run 100 m in under 8 s. If that is accepted, then the ultimate limit must lie somewhere between the current record and the 8-s mark.

It is common to read that performance improvements will become progressively smaller, but there is no sign of that in some events. Even if the rate of progress does slow, measurement methods are becoming more precise, and performance can be measured to hundredths or even thousandths of a second. This can be seen with the 100-m world record mark. When times were recorded to 0.1 s, the record was broken only five times between 1912 and 1960: with the advent of electrical timing systems and the recording of times to 0.01 s, the record has been broken 14 times since 1960. Another example of this can be seen in the results of the Oxford–Cambridge University boat race, which has been contested

Figure 1.2

(a) Linear regression equation based on male 100-m sprint world records since 1912. (b) Non-linear regression analysis applied to the same data.

(a)

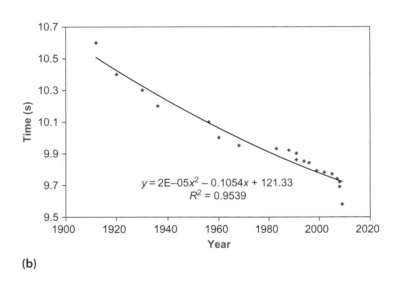

(b)

over a distance of about 4¼ miles (6.8 km) since 1829. In the 1877 race, a dead heat was declared as the judges could not separate the teams, even though it was generally agreed that Oxford had been about 6 feet (2 m) ahead when they crossed the finishing line. In the 2003 race, automatic electronic timing and slow-motion video analysis allowed the judges to declare Oxford winners by a margin of 1 foot (30 cm).

Now that more women are able to compete, performances have improved rapidly

The factors that limit human performance are of interest to scientists who study sport. The emergence of African athletes in track and marathon events and of

Figure 1.3

Evolution of world record times for the marathon for men and women since official record keeping began.

the Chinese across a range of sports has changed world sport in recent years, but the reasons for their dominance are not well understood. Some of the answers will be provided by social scientists, who look at factors such as opportunities for participation in sport and at financial and other motivations to succeed. We can see that the gradual relaxation of the social pressures that excluded most women from taking part in any form of strenuous physical activity can account for the low starting point and faster progression of women's performances (Figures 1.1, 1.3 and 1.4). With more women able to compete, and with the social barriers to hard training by women being removed—at least in most

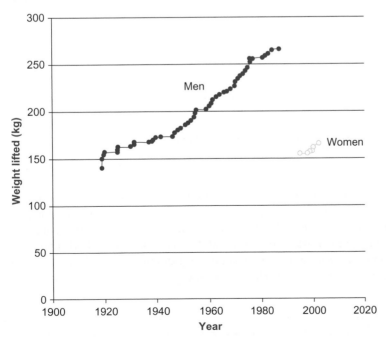

Figure 1.4

Evolution of world records for weightlifting (the clean and jerk) for males and females since official record keeping began.

societies—performances have improved rapidly as a greater fraction of the total gene pool is available. The biological sciences, however, will provide many of the answers, and the physiologist, the biochemist, the nutritionist, the molecular biologist, and others all have contributions to make to our understanding of the whole picture. In particular, each is interested in the limitation to performance, and each may have a different view of where that limitation may lie and how it can be effectively pushed back to improve performance.

Energy metabolism is a vital element in successful performance

 Leaders in the field

Otto Warburg (1883–1970) Warburg studied chemistry under Emil Fischer at the University of Berlin before he obtained a medical degree at Heidelberg in 1911. In 1918 he was appointed Professor at the Kaiser Wilhelm Institute for Biology before becoming Director of the Kaiser Wilhelm Institute for Cell Physiology. His first research with Fischer was in the polypeptide field, but he then began to work on the process of oxidation. Warburg investigated the metabolism of cells, particularly cancer cells, and in 1931 he was awarded the Nobel Prize for his 'discovery of the nature and mode of action of the respiratory enzyme.' In 1924, Warburg hypothesized that cancer, malignant growth, and tumour growth are caused by the fact that tumour cells derive most of their energy by glycolysis rather than by oxidative metabolism, and he believed that cancer should be interpreted as a mitochondrial dysfunction. In 1944, Warburg was nominated a second time for the Nobel Prize for his work on nicotinamide, the mechanism and enzymes involved in fermentation, and the discovery of flavonoids.

At the other end of the spectrum of distance covered in Olympic competition (if we exclude the 50-km walk and the triathlon), performances in the marathon have also improved dramatically in the time since the distance was standardized and records kept (Figure 1.3). The men's record is edging closer to the 2-h mark and the best women will beat almost all of the men in mixed races. In these events, strength and power are only minor components: stamina—the ability to work hard for prolonged periods without succumbing to fatigue—is the key issue. Speed may be important in a finishing sprint, but this is rare, and the speeds at the end of a race such as a marathon are still relatively slow. Physiological characteristics, such as the ability of the cardiovascular system to supply oxygen and fuels to the working muscles, and the ability to regulate body temperature to prevent overheating, are important in all endurance events. Energy metabolism, however, remains a vital element in successful performance. Carbohydrate is an especially important fuel in exercise, and the body must conserve its limited supplies of this precious resource by using fat as a fuel for energy where possible. The muscles of endurance athletes are particularly good at burning fat to provide energy, and endurance training increases this capacity still further.

As in the sprints, the performances of even the best women are mediocre by the standards of the best men, but the best women athletes would beat most (i.e. more than 99%) of the male population in open competition. It seems unlikely, however, that the best men will ever be beaten by the best women, even though the gender gap is closing. This is at least in part because of the actions of the male hormone testosterone on muscle growth and on aggression, both of which are important in strength and power events. There are also other physiological and biochemical differences between males and females that make it more likely that the best men in an event will beat the best women. There have been a few cases throughout the history of organized sport of men competing as women in order to gain an unfair advantage. Even now, though, at the time of writing, the issue of gender has again become controversial and there is a growing recognition of those individuals who are neither completely male nor completely female.

In a sport such as weightlifting, the equipment used is simple and few modifications are possible. Judging standards may have changed slightly, allowing scope for more effective lifting techniques to be accepted within the rules, but the restructuring of weight classes by the International Weightlifting Federation in 1993 and again in 1998 resulted in all records set prior to this being annulled, so tracing the development of performance becomes complicated. The official current men's clean and jerk record in the 105+ kg category stands at 263 kg by Hossein Rezazadeh of Iran, but this is still less than the 266 kg lifted by Leonid Taranenko of the Soviet Union in 1988. Figure 1.4 shows the evolution of records prior to the IWF changes, and this clearly indicates a progressively greater force-generating capacity of the muscles over time. Whether this is simply due to improved training, better technique, or other factors is not clear. Some would say that the lower level of performance today may reflect greater vigilance by those responsible for enforcing the anti-doping regulations.

> Strength, power, endurance, and skill have a physiological and biochemical basis that is mostly determined by genetic endowment and adaptation to training

These issues, which have led to improvements in performance over the years, raise questions as to what limits performance in the various sporting disciplines. Clearly, factors such as strength, power, endurance, and skill have a physiological and biochemical basis that is in part genetically determined and in part the result of adaptation to training. Many other factors will also contribute, however, including a range of psychological attributes, tactical awareness, and the ability to tolerate demanding training and competition schedules without succumbing to illness and injury. It must also be recognized that the use of banned drugs (e.g. anabolic steroids, growth hormone, erythropoietin, and stimulants) may have contributed to the improvements of records in some events.

Fatigue can be the result of an inability to supply energy as fast as the muscle is using it

The limiting factor in performance depends on the event. In strength and power events, the force-generating capacity of the muscles is critical. This depends in large part on the interactions between two protein molecules, actin and myosin, that are so arranged that they can interact to cause the ends of the muscles to be drawn towards each other—the more of these contractile proteins that are present in the muscle, the greater the force that can be generated. To do work, the muscles must be supplied with energy, and the key to energy metabolism in cells is the compound adenosine triphosphate (ATP). The faster the rate of work that the muscles are called upon to perform, the faster the rate at which ATP must be supplied. From the biochemist's perspective, fatigue is the result of an inability to supply energy (ATP) as fast as the muscle is using it.

Clearly, a greater rate of work can be achieved in a sprint than in a middle distance event, and the maximum sustainable power declines with the distance of a race (which is the same as saying that it declines with time). The muscle has various options for the generation of fresh supplies of ATP, and these strike a balance between achieving a high power output and being able to perform for prolonged periods. This book explores these various options in detail and reviews the energy sources available to muscle, the way in which they set the limit to performance in different sports events, and how they can be modified by training, diet, and other factors.

The body's biochemical processes ultimately underpin all the factors that limit performance but these processes should not be considered in isolation

If we want to improve performance in sport, we must first understand the various factors that limit performance. Although failure of ATP supply to match the demand is perhaps a fundamental cause of fatigue, the ability to generate ATP may be limited by physiological mechanisms, such as the ability of the cardiovascular system to supply oxygen and nutrients to the working muscles, so the biochemical limitations cannot be considered in isolation. In sports where skill is a key factor, the ability of the brain to control muscle actions in a coordinated way is also key: this skill involves the brain, the nerve fibres that convey signals to the muscles, and the muscles themselves, but these are also essentially biochemical processes. Even in simple activities like running and cycling, the use of pharmacological agents that affect central neurotransmitters such as serotonin and dopamine to improve performance shows that the brain has a major role to play in determining performance in some situations. Training may improve function, but it does so, at least in part, by a selective stimulation of synthesis and breakdown of proteins in the various tissues: a physiological outcome, such as an enhanced capacity for the heart to pump blood around the circulation, results from biochemical changes in the heart and the blood vessels. Nutritional interventions can improve performance, but again they do so by affecting the body's biochemistry. This may involve increasing the available fuel supply for the muscles, increasing the supply of molecules that are essential for cellular function, or modulating the changes in the synthesis of new proteins in response to training. These are only a few examples.

The appliance of science

An understanding of the scientific basis of sports performance allows better planning of training and competition, with improved performance and reduced risk of illness and injury

In the same way that sports performance has improved since record keeping began, so scientific understanding has expanded rapidly. The elite athlete or team is now supported by a range of scientific and medical specialists, including, on the science side, physiologists, psychologists, biomechanists, nutritionists, and biochemists. Medical specialists providing further support include general practitioners, cardiologists, physiotherapists, podiatrists, and others. In each of these specialities, the understanding of the scientific basis of sports performance allows better planning of training and competition, with improved performance and reduced risk of illness and injury as the key goals.

Many factors have contributed to the evolution of our understanding of the limits to performance, and the contributions of countless scientists have ensured a steady progress. Occasionally there are leaps forward due to brilliant new ideas or the introduction of a new technique that produces new information, but more often there is a slow and steady accumulation of new information. A wide range of biochemical methods is now available to the sports scientist, and the application of techniques developed for use in the biochemistry laboratory and in clinical medicine ensures that sports science can continue to progress.

This book looks at sport from a biochemical perspective, using the example of different sports events to illustrate the biochemical processes involved and to show how an understanding of these processes can shape the athlete's training and competition strategies. Each chapter begins with a list of learning objectives: these are the concepts that you can expect to have understood after reading the chapter. Each chapter ends with a list of key points that summarize some of the more important take-home messages. At the end of each chapter there is also a list of selected further reading. These are other books, journal papers, and articles from which the interested reader can find additional information related to the concepts dealt with in each chapter. Students who are unfamiliar with some of the basic concepts of chemistry and biochemistry, including the nature of atoms, molecules, and chemical bonds, acids, bases, and buffers, processes of membrane transport and cellular organelles, are advised to read Appendix 1 before reading further chapters of this book.

The abuse of science

Some athletes are willing to go to almost any lengths in the pursuit of success. For most, this means training to the limits of what their body will tolerate and sacrificing many of the pleasures that their less ambitious friends enjoy. Some athletes, however, are willing to go even further. Breaking the rules of sport is cheating, but when cheating involves the use of drugs to enhance performance there may be even greater penalties to pay. Many drugs can enhance performance, but the gains are usually short term and come at the cost of risks to health. The primary function of the anti-doping regulations is to protect the health of the athlete, with ensuring fair play a secondary consideration.

Responsibility for deciding what is against the rules and for the detection of those who take drugs lies with the World Anti-Doping Agency (WADA). Their list of doping substances and methods includes a number of agents that have important effects on the body's metabolism. Some of these, including naturally occurring hormones such as testosterone, insulin, and growth hormone, have powerful anabolic (tissue-building) actions that can aid muscle growth and repair. Synthetic steroids usually have a structure based on that of testosterone, but the structure of the parent molecule can be modified to accentuate specific anabolic or androgenic (male-like) actions or in an attempt to evade detection. So-called 'designer steroids' have been produced by chemists and pharmacologists who understand molecular structure and the actions of these drugs on human tissues. More recently, attention has begun to shift towards drugs that can provide the same effects without some or all of the unwanted side-effects. Androgen receptors (AR) are the molecular receptors located on the surface of target cells that testosterone and related steroids interact with. When the steroid molecule binds to the receptor on the cell membrane, this complex then initiates a series of intracellular events that act on the cell's nucleus to begin the process of gene transcription, i.e. to begin making more proteins. Selective androgen receptor modulators (SARMs) are compounds that are not steroids but can interact with specific steroid receptors on the cell membrane: by doing so, they can increase muscle mass and bone density while promoting fat loss and showing a lower tendency to produce unwanted side-effects. As with most drugs that are abused in sport, they are being developed for treatment of a range of diseases that were previously treated with anabolic steroids. Compounds like SARMS offer the prospect of more effective treatments for many diseases and can also be used as powerful research tools to provide new insights into the control of gene expression in tissues. They are still in the early stages of development for clinical use, but it seems likely that their abuse by athletes has begun, even before tests on their safety have been completed.

Many other categories of drugs are used by athletes and large sums of money are spent on their detection. Their use will not be described in detail here, though the study of doping agents can shed light on many aspects of exercise biochemistry.

KEY POINTS

1. Biochemistry is concerned with the study of events at the molecular and cellular level.

2. The innate level of ability and the ability to respond to training with an improvement in performance are the keys to success in sport.

3. The genetic material dictates the formation of proteins and these in turn control metabolism.

4. In sprint events improvements in performance may come from faster reaction times, better ability to generate power in the active muscles, improved resistance to fatigue, or some combination of these.

5. In most running events there has been a steady improvement in performances for as long as records have been kept and there is no obvious indication that the rate of improvement has slowed. However, there must be a limit to performance and at some point in the future there will be no further improvement.

6. With more women able to compete, and with the social barriers to hard training being removed, performances have been seen to improve rapidly as a greater fraction of the total gene pool is available.

7. Energy metabolism is a vital element in successful performance. Carbohydrate is an especially important fuel in exercise, and the body must conserve its limited supplies by using fat as a fuel for energy where possible. The muscles of endurance athletes are particularly good at burning fat to provide energy, and endurance training increases this capacity still further.

8. Factors such as strength, power, endurance, and skill have a physiological and biochemical basis that is in part genetically determined and in part the result of adaptation to training. Many other factors contribute to success in sport, however, including a range of psychological attributes, tactical awareness, and the ability to tolerate demanding training and competition schedules without succumbing to illness and injury.

9. The body's biochemical processes underpin all the factors that limit performance, but these processes should not be considered in isolation. Although ATP supply is perhaps a fundamental cause of fatigue, the ability to generate ATP may be limited by physiological mechanisms, such as the ability of the cardiovascular system to supply oxygen and nutrients to the working muscles, or by the willingness of the brain to continue.

10. The understanding of the scientific basis of sports performance allows better planning of training and competition, with improved performance and reduced risk of illness and injury as the key goals.

11. The use of prohibited drugs in sport is a breach of the rules that govern participation. Some drugs have powerful actions on metabolic processes that can enhance performance. Most also carry significant health risks.

2 The weightlifter

Learning objectives

After studying this chapter, you should be able to . . .

1. give a general description of the structure and function of skeletal muscle;

2. describe in detail the molecular composition of the myofilaments and the structure of sarcomeres;

3. understand the mechanism of muscle contraction and the determinants of muscle strength;

4. describe the biochemical and physiological characteristics of Type I and II fibre types;

5. describe the different types of muscle contraction;

6. appreciate the damaging effects of high force eccentric actions and the causes of delayed-onset muscle soreness;

7. appreciate the effects of training and ageing on muscle composition and function;

8. describe the structure and function of protein molecules;

9. understand the concept of protein turnover and the factors affecting protein synthesis and breakdown;

10. understand the mechanisms of enzyme action and the factors that can modify enzyme activity;

11. describe the role of ATP as the source of energy for muscle contraction;

12. describe the contribution of protein to energy expenditure at rest and during exercise;

13. discuss the effect of resistance exercise and feeding on protein synthesis and breakdown;

14. describe the recommendations generally given for strength athletes;

15. discuss the need for protein supplementation in athletes;

16. describe the potential health hazards of excess intake of protein;

17. discuss the effects of ingesting single amino acids;

18. discuss the actions of anabolic steroids and other compounds that are claimed to boost muscle mass and strength.

Introduction

Muscular strength is, to a large extent, determined by the size of the muscles and the ability to fully activate the muscles

Muscular strength is, to a large extent, determined by the size of the muscles (their cross-sectional area) and the ability to fully activate the muscles in a coordinated manner. Successful weightlifting requires a large muscle bulk and the ability to generate high power for a very limited period, usually less than a few seconds. Technique is also obviously important because, in competition, the weightlifter is required to demonstrate control of the posture and stance when lifting and holding the weight above the head. This chapter begins by describing skeletal muscle structure and function. Muscle proteins provide the framework of the contractile machinery, and therefore the structural and functional characteristics of these important biomolecules, including their role as enzymes, are dealt with here. Finally, the energy needs for lifting heavy weights are considered, with emphasis on the role of adenosine triphosphate (ATP) as the energy currency of the muscle (and all other) cells. Other sports also require moments of explosive power (e.g. the serve in tennis, throwing a shot-put or javelin, kicking a ball in rugby or football) and similar principles apply.

Muscle structure and function

Types of muscle

Skeletal muscle is under direct voluntary control

Muscle is one of the four primary tissue types in the body, the others being nervous, connective, and epithelial tissue. There are three forms of muscle in the body: cardiac muscle, found only in the heart; smooth muscle, located in the walls of blood vessels, airways, the gut, and the bladder; and skeletal (or striated) muscle, whose fibres link parts of the skeleton. Only skeletal muscle is under direct voluntary control and allows movement of the limbs to take place as well as the maintenance of posture.

Structure of skeletal muscle

Individual muscles are made up of many parallel muscle fibres

Skeletal muscles are separated from their surroundings by a membranous layer of connective tissue, known as the perimysium or fascia. Connective tissue also extends into the interior belly of the muscle as septa of decreasing thickness (endomysium), subdividing muscle into smaller and smaller compartments (Figure 2.1). The smallest of these is the fasciculus, which contains a number of muscle fibres bound together and to the endomysium by looser connective tissue. At both ends of the muscle this connective tissue skeleton converges to form tendons.

Tendons are tough, relatively inelastic bands of tightly packed collagenous fibres that form the connections between muscles and bones. The outer collagenous membrane of living bone (the periosteum) is continuous with the tendinous fibres. This important linkage allows contractions of the muscles to effect movements of the limbs.

Individual muscles are made up of many parallel muscle fibres that may (or may not) extend the entire length of the muscle. Inside muscle, the connective tissue also envelops the larger blood vessels and nerves that supply muscle. Almost all muscle fibres are innervated by only one nerve ending located near the middle of each muscle fibre. The specialized synapse separating the nerve and muscle cell membranes is called the motor end plate and the neurotransmitter released from the nerve ending that initiates contraction is acetylcholine. The blood vessels are generally orientated in parallel with the muscle fibres and numerous capillaries run in the spaces between the individual muscle fibres. The vasculature of muscle may constrict or dilate under nervous, hormonal, and local control to regulate blood flow. During dynamic exercise blood flow through muscle may increase

Figure 2.1

Gross and microscopic anatomy of skeletal muscle.

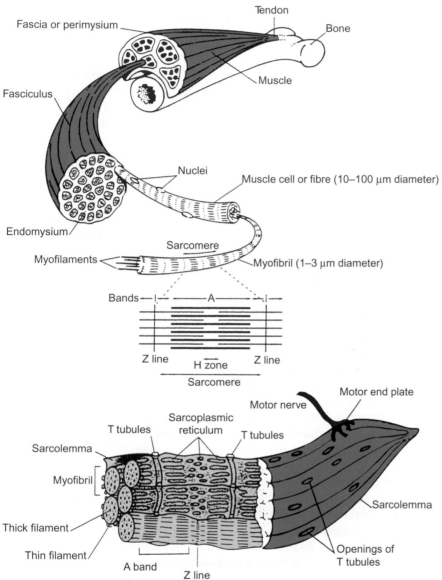

up to 100 times the resting rate. However, for the weightlifter, the contractions are of high force and, importantly, often of a static nature (called an isometric contraction). During such contractions high intramuscular pressures are generated and when the intramuscular pressure exceeds the arterial blood pressure, this effectively occludes blood flow into and out of the active muscles. Under such conditions, the energy for contraction must be obtained largely without the use of oxygen (i.e. from anaerobic metabolism).

The diameter and the cross-sectional area of the muscle fibres are influenced by training

Muscle cells are long fibres with many nuclei (from ~10 to ~3000; see Figure 2.1). They vary in length from a few millimetres to over 30 cm and are thin (10–100 μm) cylindrical threads running from end to end in the muscles. The diameter and the cross-sectional area of the muscle fibres are influenced by training. Performance of repeated bouts of heavy resistance exercise over a period of months results in muscle fibre hypertrophy: each individual muscle fibre becomes bigger. Each muscle fibre is surrounded by a homogeneous membrane (the sarcolemma) that contains collagen fibres in its outer layer that connect it with the intramuscular connective tissue elements. The inner layer of the sarcolemma is the cell membrane proper, important in production and conduction of electrical excitation along the fibre. Invaginations of the sarcolemma called the T tubules allow transmission of the action potential to the interior of the muscle fibre.

Muscle fibre ultrastructure

Under the light microscope muscle fibres have a striated appearance

The interior of the muscle fibre is filled with sarcoplasm (muscle cell cytoplasm), a red viscous fluid containing nuclei, mitochondria, myoglobin, and about 500 thread-like myofibrils of thickness 1–3 μm continuous from end to end in the muscle fibre (see Figure 2.1). The red colour is due to the presence of myoglobin, an intracellular respiratory pigment. Surrounding the myofibrils is an elaborate form of smooth endoplasmic reticulum called the sarcoplasmic reticulum. Its interconnecting membranous tubules lie in the narrow spaces between the myofibrils, surrounding and running parallel to them. Under the light microscope muscle fibres have a striated appearance. The striations are due to the unique cross-banding arrangement of the myofibrils. Dark A bands alternate with light I bands along the length of each myofibril; these are the contractile elements. As you can see from Figure 2.2, each A band is interrupted at its midsection by a lighter stripe called the H zone, which is visible only in relaxed muscle fibres. The H zone itself is bisected by a dark line called the M line (containing M-protein, myomesin, and the muscle isoform of creatine kinase). The I bands also have a midline interruption: a dark area called the Z line. A sarcomere is defined as the region between two successive Z lines in a myofibril, and is the smallest contractile unit or segment of a muscle fibre. Each myofibril is effectively a chain of sarcomeres laid end to end.

The thin filaments are composed of the proteins actin, tropomyosin, and troponin and the thick filaments contain the protein myosin

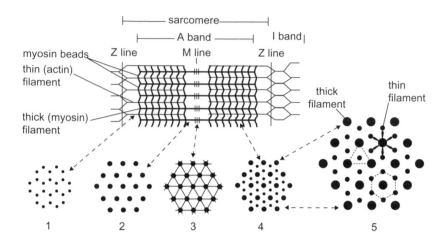

Figure 2.2

Arrangement of the thick and thin filaments in cross-section.

At the molecular level it can be seen that the banding pattern of the myofibril arises from the orderly arrangement of two types of protein filaments (myofilaments) within the sarcomeres. The thin filaments are composed of the proteins actin, tropomyosin, and troponin and extend across the I band and part way into the A band; the thick filaments contain the protein myosin and extend the entire length of the A band. The Z line is a protein sheet in the shape of a disc lying at right angles to the axis of the muscle fibre, and serves as the point of attachment of the thin filaments and also connects each myofibril to the next throughout the width of the muscle fibre. The H zone is the region in which the thick filaments are not overlapped by the thinner ones and is thus lighter when viewed microscopically than the rest of the A band. The M line in the centre of the H zone is evident as a slightly darker line because of the presence of fine strands that link together adjacent thick filaments. Serial cross-sectional views of a myofibril reveal that in areas where thick and thin filaments overlap, each thick filament is surrounded by a hexagonal arrangement of six thin filaments, and each thin filament lies within a triangle of three thick filaments (see Figure 2.2).

> The architecture of the sarcomere is maintained by several cytoskeletal proteins

The architecture of the sarcomere is maintained by several cytoskeletal proteins that keep the contractile and regulatory proteins in the correct spatial arrangement for force generation by making serial and lateral connections between adjacent myofibrils. Desmin is a protein that links adjacent Z discs in a longitudinal and lateral manner and is largely responsible for the almost perfect alignment of adjacent Z discs and A bands. At least five different proteins are thought to have a role in the anchoring of the cytoskeleton to the sarcolemma by structures called costameres. These proteins include actin, α-actinin, vinculin, talin, and dystrophin and play a vital role in force transmission to the endomysium. Within the sarcomere, actin is joined to the Z disc by a large elastic protein called α-actinin. Another large protein associated with actin is nebulin, which is thought to have a role in regulating F-actin (thin filament) length. The distance from the M line to the Z disc (about 1.0 μm) is spanned by titin, a giant protein with an elastic and also a non-elastic region that helps to maintain myosin (thick filament) alignment and support passive tension.

Calcium is released from the sarcoplasmic reticulum into the sarcoplasm and subsequently activates contraction

When calcium and ATP are present in sufficient quantities, the filaments interact to form actomyosin and the muscle shortens as they slide over each other. Electrical excitation passing as an action potential along the sarcolemma and down the T tubules leads to calcium release from the sarcoplasmic reticulum into the sarcoplasm and subsequent activation and contraction of the filament array. Excitation is initiated by the arrival of a nerve impulse at the muscle membrane via the motor end plate.

Molecular composition of the thick and thin myofilaments

The thick filaments contain mysosin and the thin filaments contain actin

Each thick filament contains about 200 myosin molecules, each of which has a rod-like tail, with two globular heads at one end that have ATPase enzyme activity. The myosin heads interact with specific sites on the thin filaments to form the so-called cross-bridges and generate the tension associated with muscle contraction. The myosin molecules in each thick filament are bundled together such that their tails form the central portion of the filament and their heads face outwards and in opposite directions to each other (see Figure 2.3). Thus, each thick filament has a relatively smooth central section and its two ends are studded with a staggered array of myosin heads. The thin filaments are composed of actin and several regulatory proteins. Globular (G) actin monomers are polymerized into long strands called fibrous (F) actin. Two F-actin strands twisted together form the backbone of each thin filament. Successive rod-shaped tropomyosin molecules spiral about the F-actin chains and help to stiffen the filament. The other main protein present in the thin filaments is troponin, which contains three subunits: troponin I binds to actin, troponin T binds to tropomyosin and troponin C can bind calcium ions (Figure 2.3).

(a) myosin molecule

(b) portion of a thick filament

(c) portion of a thin filament showing how G-actin monomers are arranged into two twisted F-actin strands

Figure 2.3

Molecular structure of thick and thin filaments.

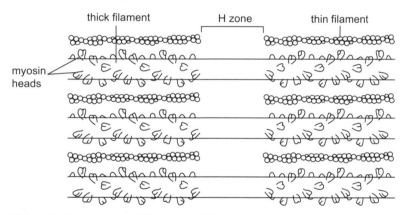

(d) longitudinal section of filaments within one sarcomere

The contractile mechanism

Each cross-bridge attaches and detaches several times during a contraction

When a muscle fibre contracts, its sarcomeres shorten, the H zone disappears and the distance between successive Z lines is reduced. The filaments themselves do

not change in length. Rather, the thin filaments slide past the thick ones so that the extent of myofilament overlap increases. Sliding of the filaments begins when the myosin heads form cross-bridges attached to active sites on the actin subunits of the thin filaments. Each cross-bridge attaches and detaches several times during a contraction, in a ratchet-like action, pulling the thin filaments towards the centre of the sarcomere. Effective shortening depends on the myosin head going through the whole cycle; the detachment phase is particularly important during shortening as myosin heads that do not detach in time interfere with, and slow down, the rate of shortening. The distance by which each sarcomere shortens is very small (about 0.5–1.0 µm), but as this event occurs in sarcomeres throughout the cell, the whole muscle fibre shortens in length considerably. If an average sarcomere is 2.5 µm in length, there will be 400 per mm, i.e. 4000 per cm.

Muscle contraction speed depends on the rate at which the myosin heads attach, rotate, and detach and on the muscle length

The speed of a muscle depends on two factors: first, the rate at which the myosin heads attach, rotate, and detach. There are various types of myosin—called myosin isoforms—that differ in their rate of reaction (as is the case with many enzymes). The main division is into slow [Type-I (one)] and fast [Type II (two)] myosin isoforms. Second, the speed of a muscle depends on the muscle length. If each sarcomere shortens by 1.0 µm, then the muscle shortens by 40% or 4 mm/cm length. If it takes one sarcomere 1s to shorten by 1.0 µm, then it follows that a 1-cm muscle will shorten at 4 mm/s, while a 10-cm muscle will shorten at 40 mm/s. You can see the relationship between muscle length and isometric force in Figure 2.4. The relationship between force and velocity of shortening is shown in Figure 2.5.

The attachment of the myosin cross-bridges requires the presence of calcium ions. In the relaxed state calcium is sequestered in the sarcoplasmic reticulum, and in the absence of calcium the myosin binding sites on actin are physically blocked by the tropomyosin rods (Figure 2.6). When calcium ions are released from the sarcoplasmic reticulum (following excitation by a nerve impulse) they bind to troponin C, causing a change in its conformation that physically moves tropomyosin away from the myosin binding sites on the underlying F-actin chain.

The power stroke of the cross-bridge cycle happens as the myosin head changes from its cocked to its bent shape

Activated or 'cocked' myosin heads now bind to the actin, and as this happens the myosin head changes from its activated configuration to its bent shape, which causes the head to pull on the thin filament, sliding it towards the centre of the sarcomere (these events are illustrated in Figure 2.7). This action represents the

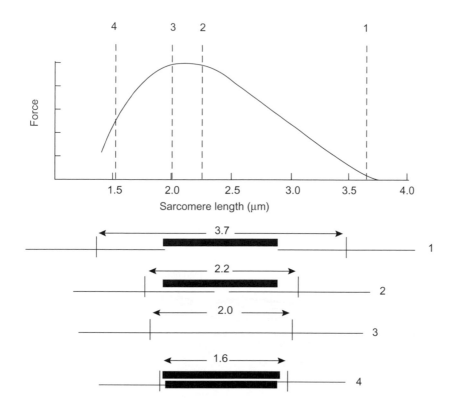

Figure 2.4

Relationship between sarcomere length and isometric force.

power stroke of the cross-bridge cycle, and simultaneously ADP and inorganic phosphate (P_i) are released from the myosin head. As a new ATP molecule binds to the myosin head at the ATPase active site, the myosin cross-bridge detaches from the actin. Hydrolysis of the ATP to ADP and P_i by the ATPase provides the

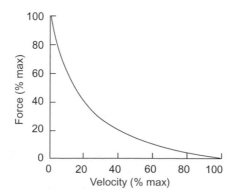

Figure 2.5

Relationship between force and velocity of shortening (both variables are expressed as the percentage of the maximum value).

Figure 2.6

Each troponin molecule contains three subunits (C, I, and T) with binding sites for calcium (C), actin (I), and tropomyosin (T). In the relaxed state the presence of tropomyosin physically prevents the interaction of actin and myosin. When calcium ions bind to the C-subunits of troponin molecules, a shape change occurs that pushes the tropomyosin rods away from the actin biding sites, allowing interaction of the actin filaments and myosin filaments to take place.

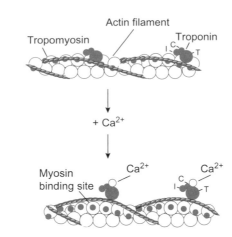

Figure 2.7

Sequence of events in cross-bridge formation.

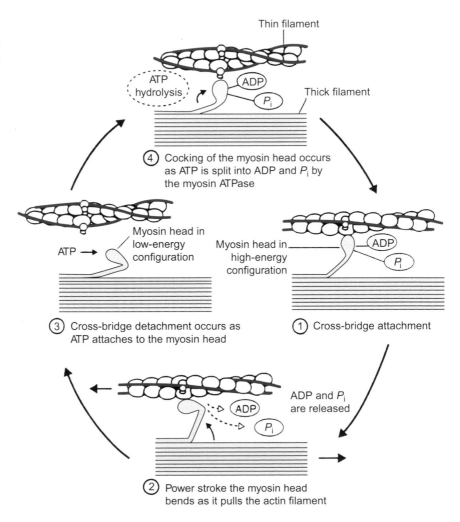

energy required to return the myosin to its activated 'cocked' state, empowering it with the potential energy needed for the next cross-bridge attachment–power stroke sequence. While the myosin is in the activated state, the ADP and P_i remain attached to the myosin head. Now, the myosin head can attach to another actin unit further along the thin filament, and the cycle of attachment, power stroke, detachment, and activation of myosin is repeated. Sliding of the filaments in this manner continues as long as calcium is present (at a concentration in excess of 10 µM) in the sarcoplasm. Removal and sequestration of the calcium by the ATP-dependent calcium pump (ATPase) of the sarcoplasmic reticulum restores the tropomyosin inhibition of cross-bridge formation and the muscle fibre relaxes.

Control of contraction

Transmission of the action potential to the sites where the T tubules adjoin the SR causes the release of calcium ions

For a muscle fibre to contract, a nerve impulse from the motor nerve that innervates the muscle fibre must result in the propagation of an action potential along the sarcolemma. An action potential arriving at the motor end plate causes release of the neurotransmitter acetylcholine (ACh), which traverses the specialized synapse between the nerve ending and the muscle fibre (the neuromuscular junction) and attaches to acetylcholine receptors on the sarcolemma. This causes the opening of sodium channels, resulting in sodium influx down its concentration gradient into the muscle fibre, depolarization of the membrane and hence the initiation of an action potential that is then conducted along the muscle fibre sarcolemma in both directions and down the T tubules and results in the full contraction of the muscle fibre (see Figure 2.8). Transmission of the action potential to the sites where the T tubules adjoin the sarcoplasmic reticulum (SR) causes the latter to release calcium (calcium ion channels temporarily open) and the sarcoplasmic free calcium concentration rises to above 10 µM, allowing cross-bridge formation to be initiated as described previously. The continuously active calcium pump returns the calcium to the sarcoplasmic reticulum (usually within about 30 ms) and when the calcium concentration in the sarcoplasm becomes too low, the inhibition of tropomyosin is re-established. This sequence of events is repeated when another motor nerve impulse arrives at the motor end plate. When impulse frequency is high, calcium ions continue to be released from the sarcoplasmic reticulum and calcium concentration in the sarcoplasm surrounding the myofilaments increases greatly. In this situation, the muscle fibres do not completely relax between successive stimuli and contraction will be stronger and more sustained (up to a point) until nervous stimulation ceases.

Figure 2.8

Sequence of events in excitation–contraction coupling.

Motor units

Groups of muscle fibres within the muscle are functionally united

A single motor neuron forms synapses to many individual muscle fibres and all will respond when the nerve fibre is activated. Thus, groups of muscle fibres (of the same fibre type) within the muscle are functionally united through their connection with the same motor neuron. Each group is called a motor unit (as illustrated in Figure 2.9). Motor units vary in the numbers of individual fibres they comprise: some contain as few as 50 fibres, others up to 1700 fibres. Muscles that perform finely graduated movements (e.g. of the eyes and hands) tend to have small motor units; those involved in coarser movements of larger masses (e.g. the arms and legs) have larger motor units.

An impulse travelling down the motor nerve axon causes depolarization under its motor end plates. The muscle fibres in the unit will either (a) not respond or (b) respond with a conducted action potential along the muscle fibres, followed by simultaneous contraction of all the fibres.

Figure 2.9

Innervation of muscle fibres by motor neurons, showing a single motor unit in which a single motor neuron forms neuromuscular junctions with a group of muscle fibres (all of which are of the same fibre type).

A fully fused maximum tetanus has a much higher tension than a single twitch

As you can see from Figure 2.10, a muscle fibre responds to a single stimulus (of sufficient intensity) by a twitch contraction (and relaxation) lasting about 30 ms. This is because a single action potential will cause the release of a quantity of calcium into the interior of the muscle fibre, causing a brief interaction of actin and myosin and the generation of force—the twitch. With low-frequency repeated stimulation the muscle responds with a series of separate twitches, but as the frequency of stimulation increases, the twitches begin to overlap and the forces summate, giving first an unfused tetanus and then a fully fused maximum tetanus (see Figure 2.10) with a much higher tension than a single twitch. The maximum force is generated when the stimulation is sufficiently fast to release enough calcium from the sarcoplasmic reticulum to saturate all the troponin binding sites.

The normal frequency of stimulation varies from about 5 Hz, which produces weak contractions, up to about 70 Hz, in the production of strong contractions. An increase in the tension generated by a whole muscle contraction can be effected by (a) increasing the frequency of stimulation of active motor units and (b) recruiting an increasing number of motor units and hence activating a larger number of muscle fibres.

The force of a maximal contraction is determined by how many motor units are simultaneously activated and the cross-sectional area of muscle involved

During exercise of fixed intensity some motor units drop out when fatigued but their contribution to force generation will be replaced immediately by another until all motor units have been recruited. During maximal exercise, the initial full recruitment of (almost) all motor units is followed by a gradual loss of some units as fatigue ensues. The force of a maximum voluntary contraction is thus determined by how many motor units are simultaneously activated and the

Figure 2.10

Isometric muscle contractions produced by a single stimulus followed by trains of stimuli, first at 10 Hz, producing an unfused tetanus, then at 100 Hz, producing a fused tetanus. Note that the smoothness of the contraction increases with stimulation frequency, as does the peak force attained.

(a) Sarcomeres in series

Tendon ◄——————   ——————► Tendon

(b) Sarcomeres in parallel

Tendon ◄——————   ——————► Tendon

Figure 2.11

The maximum force generated by two sarcomeres in parallel (b) is twice that of two sarcomeres in series (a). In contrast, the maximum velocity of shortening is twice as fast for two sarcomeres in series.

cross-sectional area of muscle involved. It is the number of sarcomeres in parallel (i.e. the cross-sectional area of the muscle) that determines the strength of a muscle (see Figure 2.11). There is some evidence that sedentary individuals do not fully recruit all available motor units during maximal voluntary contractions and that at least some (and probably most) of the increase in strength during the early stages of a resistance training programme is brought about by neuromuscular adaptations that allow a fuller and better coordinated activation of available motor units. It usually takes several weeks or months of weight-training before any noticeable increase in muscle size (hypertrophy) occurs.

Fibre types

The proportions of the different fibre types in a muscle largely determine its capacity for power or endurance

The existence of different fibre types in skeletal muscle is readily apparent and has long been recognized; the detailed physiological and biochemical bases for these differences and their functional significance have, however, only been established more recently. Much of the impetus for these investigations has come from the realization that success in athletic events that require either the ability to generate a high-power output or great endurance is dependent in large part on the proportions of the different fibre types that are present in the muscles. The muscle fibres are, however, extremely plastic, and although the fibre-type distribution is genetically determined, and not easily altered, an appropriate training programme will have a major effect on the metabolic potential of the muscle, irrespective of the fibre types present.

The major functional characteristic that differentiates between fibre types is the speed of contraction and relaxation

The original basis for classification of muscle fibre types as red, white, or intermediate was that applied to whole animal muscles by gross visual inspection. A much clearer distinction can be made in some animals (e.g. fish, chicken) than in man. The major functional characteristic that differentiates between fibre types, however, is the speed of contraction and relaxation. Slow-twitch fibres have a long time to peak tension (about 80–100 ms for fibres from human muscle) and also a long half-relaxation time. In human fast-twitch fibres, the time to peak tension is about 40 ms, and the relaxation time is correspondingly shorter. The two fibre types form distinct groups with no overlap in these contractile properties. Muscles that are made up of predominantly slow or fast fibres have different shaped twitches and the stimulation frequency at which tetanic fusion occurs is different. You can see these differences by comparing the diagrams in Figure 2.12.

Fibre-type classification is usually based on histochemical staining

Because of the obvious difficulties in obtaining tissue for measurement of contractile properties, and the relative ease with which small samples of muscle can be obtained by needle biopsy, fibre-type classification is usually based on histochemical staining of serial cross-sections. On this basis, human muscle fibres are commonly divided into three major kinds: Types I, IIa, and IIX. These are

Figure 2.12

Isometric force generated by increasing stimulation frequencies in a slow muscle (soleus) and a fast muscle (extensor digitorum longus, EDL).

Biopsy. A biopsy is simply a sample of living tissue that is removed from the body for examination. It can be analysed in a number of ways, depending on what the investigator is looking for. Muscle samples, for example, can be fixed, mounted, and cut into thin sections for examination under the microscope after suitable staining techniques have been applied. Alternatively, the samples can be used for chemical analysis to determine, for example, the content of glycogen and lipid, high-energy phosphagens, or metabolic intermediates. The needle biopsy technique allows repeated removal of small samples, allowing the time course of changes in metabolism to be established. The technique came into widespread use in the 1960s and led to rapid increases in understanding of patterns of glycogen use in muscle. A major advance was the use of micromethods that allowed single muscle fibres to be dissected out of biopsy samples, stained for identification of fibre type, and analysed for their content of glycogen and high-energy phosphates. The needle biopsy technique has occasionally been applied to obtain liver samples, but it is difficult to stop the profuse bleeding that tends to occur, so such studies are seldom if ever done today on healthy volunteers, although this remains, as will all biopsy techniques, a routine diagnostic tool in clinical practice.

analogous to the muscle fibres from rodents, which have been classified on the basis of their directly determined functional properties as slow twitch (Type-I), fast twitch-fatigue resistant (Type-IIa), and fast twitch-fatiguable fibres (Type-IIb), respectively. The myosin of the different fibre types exists in different molecular forms (isoforms), and the myofibrillar ATPase activity of the different fibre types displays differential pH sensitivity; this provides the basis for the differential histochemical staining of the fibre types. The myosin ATPase of the Type-II fibres is inactivated at low pH (less than about pH 4.5) whereas the ATPase activity of Type-I fibres is unaffected. Above about pH 9 the situation is reversed, and Type-II myosin ATPase activity is stable, while Type-I myosin ATPase is inactivated. By incubation at different pH values close to the lower end of this range prior to histochemical staining for myosin ATPase activity, two distinct subtypes of the Type-II fibres can be distinguished: the myosin ATPase of the Type-IIa fibres is inactivated at pH 4.6–4.8, whereas that of the Type-IIX fibres is maintained. It is sometimes possible to detect a Type-IIc fibre type by differential pre-incubation, although these do not normally account for more than 1% of the total fibre number in human muscle. It is becoming increasingly common to use gel electrophoresis to distinguish between different fibre types based on the presence of different isoforms of the myosin heavy-chain proteins. There is a very good correlation between the myosin heavy-chain composition of a muscle and the maximum rate at which it can develop force when stimulated. Note that in human limb muscles the myosin isoform IIX is expressed, not the IIb found in small mammals such as rats and mice. The IIX isoform is a slightly slower version than the IIb isoform. Human muscles do possess the gene for IIb, but this only appears to be expressed in the muscles of the eye and larynx where particularly fast contractions are required. The physiological and biochemical characteristics of the three major fibre types are summarized in Tables 2.1a and b.

Characteristic	Type I	Type IIa	Type IIX
Nomenclature	Slow, Red	Fast, Red	Fast, White
	Fatigue resistant	Fatigue resistant	Fatiguable
Motor neuron size	Small	Large	Large
Recruitment frequency	Low	Medium	High
Contraction speed	Slow	Fast	Fast
Relaxation speed	Slow	Fast	Fast
Maximum power output	Low	High	High
Endurance	High	Medium	Low
Capillary density	High	Medium	Low
Mitochondrial density	High	Medium	Low
Metabolic character	Oxidative	Intermediate	Glycolytic
Myoglobin content	High	Medium	Low
Glycolytic enzyme activity	Low	High	High
Oxidative enzyme activity	High	High	Low
Glycogen content	Low	High	High
Triglyceride content	High	Medium	Low
Phosphocreatine content	Low	High	High
Myosin ATPase activity	Low	High	High
ATPase activity at pH 10.3	0	High	High
ATPase activity at pH 10.3 with pre-exposure to pH 4.6	0	0	High

Table 2.1a Characteristics of human muscle fibre types and qualitative differences.

Characteristic	Type I	Type IIa	Type IIX
Capillary density	1.0	0.8	0.6
Mitochondrial density	1.0	0.7	0.4
Myoglobin content	1.0	0.6	0.3
Phosphorylase activity	1.0	2.1	3.1
PFK activity	1.0	1.8	2.3
Citrate synthase activity	1.0	0.8	0.6
SDH activity	1.0	0.7	0.4
Glycogen content	1.0	1.3	1.5
Triacylglycerol content	1.0	0.4	0.2
Phosphocreatine content	1.0	1.2	1.2
Myosin ATPase activity	1.0	>2	>2

Table 2.1b Biochemical characteristics of human muscle fibre types. Values of metabolic characteristics of Type-II fibres are shown relative to those found in Type-I fibres.

PFK, phosphofructokinase; SDH, succinate dehydrogenase.

Type-I fibres are red cells that contract relatively slowly and possess a high oxidative capacity

Type-I fibres are red cells that contain relatively slow-acting myosin ATPases and hence contract slowly. The red colour is due to the presence of myoglobin, an intracellular respiratory pigment, capable of binding oxygen and only releasing it at very low partial pressures (as are found in the proximity of the mitochondria). Type-I fibres have numerous mitochondria, mostly located close to the periphery of the fibre, near to the blood capillaries that provide a rich supply of oxygen and nutrients. These fibres possess a relatively high activity of oxidative enzymes (Table 2.2) and hence have a high capacity for oxidative metabolism. Type-I fibres are extremely fatigue resistant and specialized for the performance of repeated strong contractions over prolonged periods.

Type-II fibres are pale cells that contract relatively fast and possess a high glycolytic capacity

Type-IIX fibres are much paler than Type-I because they contain little myoglobin; in strength-trained individuals they tend to be larger in diameter than Type-I

Table 2.2 Activities of some glycolytic and oxidative enzymes in different fibre types in human skeletal muscle (μmol/min/gww).

Enzyme	Type I	Type IIa	Type IIX
Phosphorylase	2.8	5.8	8.8
Phosphofructokinase	7.5	13.7	17.5
Succinate dehydrogenase	7.1	4.8	2.5
Citrate synthase	10.8	8.6	6.5

fibres although any such differences in fibre size are dependent to a large extent on patterns of habitual activity. They possess rapidly acting myosin ATPases and so their contraction (and relaxation) time is relatively fast and consequently they have about a threefold greater maximum power output than the Type-I fibres. They have few mitochondria and a poorer capillary supply but greater glycogen and phosphocreatine stores compared with the Type-I fibres. A relatively high activity of glycogenolytic and glycolytic enzymes (Table 2.2) endows Type-IIX fibres with a high capacity for rapid (but relatively short-lived) ATP production in the absence of oxygen (anaerobic capacity). Hence, these fibres are best suited for delivering rapid, powerful contractions for brief periods but they are known to fatigue rapidly.

Type-IIa fibres are red cells whose metabolic and physiological characteristics lie between the extreme properties of the other two fibre types. They contain fast-acting myosin ATPases like the Type-IIX fibres, but have an oxidative capacity more akin to that of the Type-I fibres.

During most forms of movement, there appears to be an orderly hierarchy of motor-unit recruitment from Type I to Type IIa to Type IIX

Associated with the differences in contractile speed and in the metabolic profile of the major fibre types are differences in the motor neurons that innervate the fibres: Type-I fibres are supplied by small-diameter neurons that have a slow conduction velocity and a low threshold of activation, whereas Type-II fibres are innervated by large-diameter, fast-conducting neurons that have a relatively high activation threshold. The differences in activation threshold of the motor neurons supplying the different fibre types determine the order in which fibres are recruited during exercise, and this in turn determines the metabolic response to exercise. During most forms of movement there appears to be an orderly hierarchy of motor-unit recruitment based on size, which roughly corresponds with a progression from Type I to Type IIa to Type IIX. It follows that during light exercise mostly Type-I fibres will be recruited, during moderate exercise both Type-I and Type-IIa, and during more severe exercise all fibre types will contribute to force production (as illustrated in Figure 2.13).

Figure 2.13

The ramp-like recruitment of Type-I (slow-twitch) and Type-II (fast-twitch) muscle fibres during exercise of increasing intensity. Note that during the highest intensities of exercise, all fibre types are recruited.

Large motor units tend to be composed of fast fibres, small units of slow fibres, and the motor neurons supplying them tend to be large and small, respectively. Small, slow motor units tend to be used for small, slow movements, as required in adjusting posture and maintaining balance, while the large, fast motor units are recruited for the occasional large, rapid contraction, such as jumping, lifting, throwing, sprinting, etc. The other major difference between the fast and slow motor units is in their fatiguability: slow units fatigue slowly, fast units fatigue rapidly.

Type-II fibres are not always larger than Type-I fibres

A common misconception, unfortunately repeated in many textbooks, is that in humans the Type-II fibres are always larger (i.e. have a greater diameter) than the Type-I fibres. While this is true in small mammals such as the rat, it is certainly not the case in untrained humans. Among the general sedentary population, in any given muscle the fibres of all types are of similar diameter; that is until old age, when Type-II fibres atrophy at a faster rate. In trained adults, the fibres that regularly do the most work are usually the largest. Thus, for an endurance athlete, the Type-I fibres in the leg muscles are likely to be of greater diameter than the majority of the Type-II fibres. By contrast, for the weightlifter, whose efforts involve generating very high forces, the Type-II fibres will be larger.

Muscles in the body contain a mixture of different fibre types

Whole muscles in the body contain a mixture of these three different fibre types, although the proportions in which they are found differ substantially between different muscles and can also differ between different individuals. For example, muscles involved in maintaining posture (e.g. soleus in the leg) have a high proportion (usually >70%) of Type-I fibres, which is in keeping with their function in maintaining prolonged, chronic, but relatively weak, contractions. Fast Type-II fibres, however, predominate in muscles where rapid movements are required (e.g. in the muscles of the hand and the eye). Other muscles, such as the quadriceps group in the leg, contain a variable mixture of fibre types. The fibre-type composition in such muscles is a genetically determined attribute and does not appear to be pliable to a significant degree by training. Weightlifters tend to have more Type-II than Type-I fibres, a characteristic shared by elite field event athletes and sprinters. The explanation for this may be, in part, that lifting heavy weights involves accelerating the load sharply, to carry it past unfavourable limb angles. Probably more important, however, is the fact that in response to heavy resistance training, the Type-II fibres attain a larger maximal fibre diameter than Type-I fibres and so can produce more force.

Types of contraction

Muscle contractions may be isometric, concentric, or eccentric

Skeletal muscle can perform three different types of contraction: isometric, where the muscle length remains constant; concentric, where the muscle shortens, power is generated, and work is done by the muscle, and eccentric, where the muscle is lengthened in the active state, power is absorbed, and work is done on the muscle. The weightlifter will perform all three types of muscle action: concentric in the lifting of a weight; isometric in holding a weight in a fixed position; and eccentric during the action of lowering a weight (see Figure 2.14). In addition to these pure types of contraction there are many situations where combinations of contractions occur, the most important of which may be the stretch–shortening cycle in which an eccentric action immediately precedes a concentric contraction. The preservation of high force generation in the initial eccentric phase maintains the amount of elastic energy stored. During the subsequent power-generating concentric phase the contribution of the recovered elastic energy offsets the effect of fatigue when compared to a pure concentric contraction.

Eccentric exercise-induced muscle damage

Exercise with a substantial eccentric component can cause muscle damage and delayed-onset muscle soreness

Repeated performance of high-force eccentric actions can cause muscle damage and induces temporary muscle soreness and stiffness, an effect that is usually first felt about 6–12 h after the eccentric exercise bout and persists for several days. The role of different fibre types in the generation of eccentric force is not

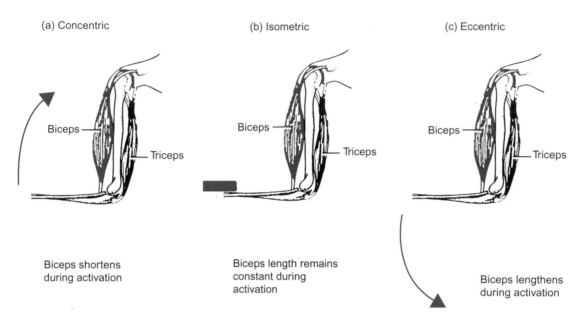

(a) Concentric

Biceps

Triceps

Biceps shortens
during activation

(b) Isometric

Biceps

Triceps

Biceps length remains
constant during
activation

(c) Eccentric

Biceps

Triceps

Biceps lengthens
during activation

Figure 2.14

Types of muscle contraction
illustrated using the biceps muscle
of the upper arm: (a) concentric;
(b) isometric; (c) eccentric.

well understood, although a fairly small number of fibres seem to be activated
to generate eccentric as compared with concentric force. There is also some
evidence that Type-II fibres may be selectively recruited during eccentric muscle
actions. This high load on relatively few fibres may cause localized ruptures of the
fibres, resulting in an inflammatory response with an accompanying development of swelling and pain. That muscle fibres are in fact damaged is evidenced
by the appearance of high levels of intramuscular enzymes in the blood in the
days following the eccentric exercise bout, and also by histological evidence of
sarcomere disruption and Z-line streaming in the affected muscles. Sarcomeres
along a muscle fibre are not uniform in length. Even at rest, some sarcomeres are
stretched more than others. It seems that when a muscle fibre is stretched during
eccentric exercise, the longer sarcomeres reach a yield point, where they pop
out of alignment. As sarcomeres get longer and longer they are able to generate
less and less force and so become easier and easier to stretch. When a popped
sarcomere relaxes, the filaments usually fit together again so that everything
returns to normal, but sometimes the filaments remain disrupted and the damage can spread to the adjacent sarcomeres during the next eccentric contraction.
Eventually the whole muscle fibre can die, leading to a weakening of the whole
muscle and pain as the white blood cells infiltrate the tissue to clear up the damaged tissue. The body responds to this damage by building a new muscle fibre
that has more sarcomeres than before. Rats that were made to run downhill on
a treadmill developed 15% more sarcomeres in their thigh muscles than rats that

ran uphill or did not exercise. After this remodelling, each sarcomere does not have to stretch so far during subsequent eccentric actions and therefore fewer sarcomeres reach the yield point where they pop. This protective adaptation occurs after a single bout of eccentric exercise and is effective in minimizing tissue damage and soreness with further bouts of eccentric exercise for up to about 10 weeks.

Adaptive capacity of skeletal muscle

Skeletal muscle possesses a considerable capacity to adapt in response to different patterns of activity or disuse

Skeletal muscle is a remarkably plastic tissue: it possesses a considerable capacity to adapt in response to different patterns of activity or disuse. Adaptation can take the form of alteration in muscle size, fibre composition, metabolic capacity, and capillary density. Long-term heavy resistance training results in muscle hypertrophy and gains in strength. Improvements in strength are important as this results in the muscle working at a lower fraction of its maximum capability for force output when required to perform work. Increases in muscle mass mean that more muscle tissue is available to perform work, resulting in a greater peak power output and also a greater total capacity of the anaerobic energy-producing systems. The penalty is the extra weight to be carried, which becomes an important factor in longer events. A more detailed consideration of the adaptations to strength training is given in Chapter 8. However, it is important to note that the gradual muscle hypertrophy that occurs with high-resistance, slow-velocity training is not achieved by increasing the number of muscle fibres, but by an increase in the diameter of the muscle fibres, which may include both Type-I and Type-II fibres. Strength training does not modify the fibre-type composition of skeletal muscle. Bodybuilders and weightlifters have fibre distributions in their muscles that are within the range of non-athletes.

Our muscles get weaker as we get older

Ageing has significant effects on muscle size and function. Maximum strength of men and women is generally attained between 20 and 30 years of age. By the age of 70, on average, the muscles are about 30% weaker. Reduced muscle mass is a primary factor in the age-associated loss of strength and may be due to a reduced muscle fibre size, particularly in the Type-II fibres. There may also be a reduction in the total number of muscle fibres, caused by a loss of motor neurons in the elderly. Innervation of muscle fibres is required for their maintenance (possibly due to the chronic production of nerve-derived growth factors) and denervation leads to muscle fibre atrophy and eventual replacement by connective tissue. The age-associated loss of muscle mass may be caused by ageing itself, or it may be a result of the reduced level of physical activity and hence the removal of the training stimulus, or it may be a combination of both of these factors. However, it is clear that muscle in the elderly still possesses the capability to adapt in response to strength training, and that significant improvements in physiological, structural, and performance characteristics can be achieved with vigorous resistance-training programmes. As with younger adults, the frequency, intensity, and duration of exercise are crucial in determining the extent of training adaptations.

Proteins: structural and functional characteristics

A tissue's protein composition determines its metabolic and functional capabilities

Proteins provide the structural basis of all tissues and organs, and it is largely the protein content of these tissues that give them their recognizable shape. More importantly, perhaps, the proteins present in the different tissues confer on each tissue its metabolic and functional capabilities. The presence of actin and myosin molecules in muscle, for example, confers the ability of the tissue to contract and the presence of specific enzymes determines whether or not a tissue can carry out a particular function. The amount of enzyme activity present determines how fast that process can proceed. Proteins and amino acids also constitute, or act as precursors for, many of the body's hormones, regulatory peptides, and neurotransmitters as well as acting as the receptors for these signalling systems and fulfilling a variety of other functions.

Amino acids

The basic structure of all amino acids consists of an amine group and a carboxyl group attached to a single carbon atom

Proteins are composed of long chains of amino acids in a linear sequence, linked together by peptide bonds. A total of 20 different amino acids are present in the body, either linked together in short chains to form peptides or linked in longer more complex structures to form polypeptides or proteins. All amino acids contain carbon, hydrogen, and oxygen, as do carbohydrates and lipids, but all also contain nitrogen; two of the amino acids (cysteine and methionine) also contain sulphur.

The basic structure of all amino acids consists of an amine ($-NH_2$) group and a carboxyl ($-COOH$) group attached to a single carbon atom; also present is an organic side chain, and the structure of these different side chains gives

Figure 2.15

General structure of amino acids, where R is one of 20 possible side chains.

the amino acids their characteristic structure (Figure 2.15): at physiological pH, these mostly exist in the ionized form as NH_3^+ and COO^-. The essential amino acids cannot be synthesized by man, and must be present in the diet, but all others can be synthesized. The individual amino acids, together with the three-letter abbreviations by which they are normally identified, are listed in Table 2.3, and the chemical structure is shown in Figure 2.16. All amino acids in animal tissues—with the exception of the simplest (glycine), which is not optically active—occur in the L-enantiomorph form: that is, they are configured such that they rotate polarized light in a particular direction (Figure 2.17). This property in itself is of no particular significance, but it does reflect the shape of the molecule, which determines its ability to interact with other molecules. This in turn determines the folding of the amino acid chains to form complex protein structures with active sites that can recognize other molecules and act as catalysts.

The concentrations of free amino acids in muscle and in the blood differ

Muscle contains all the naturally occurring amino acids. The most abundant amino acids in muscle protein are the branched-chain amino acids (BCAAs) leucine, valine, and isoleucine, which together account for about 20% of the total amino acids found in muscle protein. The most abundant free amino acid in muscle is glutamine (its concentration is about 20 mmol/kgww) and it is also the most abundant free amino acid in plasma where its concentration is about 0.7 mmol/l. The concentrations of free amino acids in muscle and in the blood clearly differ, suggesting that it is important to maintain these concentration gradients. Because different amino acids have different concentration gradients, different transporters move individual or groups of amino acids across membranes. Amino acid transporters are membrane-bound proteins that recognize specific amino acid shapes and chemical properties (e.g. neutral, basic, or anionic). The transporters are divided into sodium-dependent and sodium-independent carriers. Generally, the sodium-dependent transporters maintain a larger gradient than the sodium-independent transporters. It should be noted that all of these transporters are facilitative and usually coupled to co-transport with sodium and thus are not directly energy dependent. To date, some transporter proteins have been identified, but many more are yet to be discovered.

Name	Abbreviation	Essential
Alanine	Ala	No
Arginine	Arg	No
Asparagine	Asn	No
Aspartate	Asp	No
Cysteine	Cys	No
Glutamate	Glu	No
Glutamine	Gln	No
Glycine	Gly	No
Histidine	His	Yes
Isoleucine	Ile	Yes
Leucine	Leu	Yes
Lysine	Lys	Yes
Methionine	Met	Yes
Phenylalanine	Phe	Yes
Proline	Pro	No
Serine	Ser	No
Threonine	Thr	Yes
Tryptophan	Trp	Yes
Tyrosine	Tyr	No
Valine	Val	Yes

Table 2.3 Amino acids and their abbreviations.

Amino acids are involved in a wide variety of biochemical and physiological processes, some of which are common to all and some of which are highly specific to certain amino acids. Amino acids are constantly incorporated into proteins (protein synthesis), and proteins are constantly broken down (protein breakdown or degradation). This constant turnover of proteins is summarized in Figure 2.18. The vast majority of the amino acids in the body are incorporated into tissue proteins, but a small pool of free amino acids also exists (about 120 g of free amino acids are present in the skeletal muscle of an adult). Amino acids are constantly extracted from the free amino acid pool for synthesis of various

Figure 2.16

The structure of the amino acids found in the human body, with classification according to the nature of the side chain (R group).

Glycine

$$H.CH.COO^-$$
$$|$$
$$NH_3^+$$

Alanine

$$CH_3CH.COO^-$$
$$|$$
$$NH_3^+$$

Valine

$$CH_3$$
$$\diagdown$$
$$CH.CH.COO^-$$
$$CH_3 \diagup \quad |$$
$$NH_3^+$$

Leucine

$$CH_3$$
$$\diagdown$$
$$CH.CH_2CH.COO^-$$
$$CH_3 \diagup \qquad |$$
$$NH_3^+$$

Isoleucine

$$CH_3CH_2$$
$$\diagdown$$
$$CH.CH.COO^-$$
$$CH_3 \diagup \quad |$$
$$NH_3^+$$

Serine

$$HO.CH_2CH.COO^-$$
$$|$$
$$NH_3^+$$

Threonine

$$CH_3CH.CH.COO^-$$
$$| \quad |$$
$$OH \ NH_3^+$$

Lysine

$$^+H_3N.CH_2CH_2CH_2CH_2CH.COO^-$$
$$|$$
$$NH_3^+$$

Arginine

$$^+H_2N$$
$$\diagdown$$
$$C.NH.CH_2CH_2CH_2CH.COO^-$$
$$H_2N \diagup \qquad\qquad\qquad |$$
$$NH_3^+$$

Histidine

$$HC = C.CH_2CH.COO^-$$
$$| \qquad | \qquad |$$
$$^+HN \quad NH \quad NH_3^+$$
$$\diagdown \qquad \diagup$$
$$C$$
$$|$$
$$H$$

Aspartate

$$^-OOC.CH_2CH.COO^-$$
$$|$$
$$NH_3^+$$

Glutamate

$$^-OOC.CH_2CH_2CH.COO^-$$
$$|$$
$$NH_3^+$$

Asparagine

$$H_2NOC.CH_2CH.COO^-$$
$$|$$
$$NH_3^+$$

Glutamine

$$H_2NOC.CH_2CH_2CH.COO^-$$
$$|$$
$$NH_3^+$$

Cysteine

$$HS.CH_2CH.COO^-$$
$$|$$
$$NH_3^+$$

Methionine

$$CH_3S.CH_2CH_2CH.COO^-$$
$$|$$
$$NH_3^+$$

Phenylalanine

Tyrosine

Tryptophan

Proline

Figure 2.16

Continued

proteins and peptides, and breakdown of protein (protein degradation) makes amino acids available for the free amino acid pool. Although adults are in something close to net protein balance, salvage of the amino acids from the breakdown of proteins is not perfect so there is a need for a constant input of new amino acids from the diet.

Amino acids act as the precursors for a number of other compounds

Amino acids are used for the synthesis of amino acid-like compounds. A list of the most important products is provided in Table 2.4. The amino acids glutamate, tyrosine, and tryptophan, for example, are precursors of neurotransmitters. Glutamate is a special amino acid in this respect because it is not only a precursor of neurotransmitters but is also a neurotransmitter itself. Tyrosine is the precursor of catecholamines (dopamine, adrenaline, and noradrenaline) and tryptophan is the precursor of serotonin (5-hydroxytryptamine).

L-alanine   D-alanine

Figure 2.17

D and L forms of alanine. All amino acids with the exception of glycine are optically active: the L-form is the biologically important one.

Figure 2.18

Protein metabolism in a 70-kg human. Amino acids enter the free amino acid pool from the diet (absorption) or from breakdown of protein. Amino acids leave the free amino acid pool for protein synthesis or oxidation to CO_2 and H_2O.

Table 2.4 Products synthesized from amino acids.

Product	Synthesized from	Product	Synthesized from
Creatine	Arginine Glycine Methionine	Pyrimidines	Aspartate Glutamine
Glutathione	Cysteine Taurine Glutamine	Histamine Carnitine	Histidine Lysine
Neurotransmitters	Glutamate Tyrosine Tryptophan	Choline Serine	Methionine
Purines	Aspartate Glutamine Glycine	T3, T4 Adrenaline Noradrenaline Dopamine	Tyrosine

Amino acids are also precursors of creatine, carnitine, and various other important compounds found in cells.

Protein structure

Proteins are formed as a linear sequence of amino acids

Proteins are formed as a linear sequence of amino acids formed by a series of condensation reactions involving the carboxyl and amino groups of adjacent amino acids: the resulting chemical bond is known as a peptide bond (see Figure 2.19). Protein synthesis and its control are dealt with in Chapter 7. Here, we will only mention that synthesis of proteins is controlled by the genetic information contained

$$H_2N-\underset{\underset{H}{|}}{\overset{\overset{\boxed{R}}{|}}{C}}-\underset{\underset{}{}}{\overset{\overset{O}{\parallel}}{C}}-OH + H-\underset{\underset{H}{|}}{\overset{\overset{\boxed{R'}}{|}}{N}}-C-COOH \longrightarrow H_2N-\underset{\underset{H}{|}}{\overset{\overset{\boxed{R}}{|}}{C}}-\underset{\underset{}{}}{\overset{\overset{O}{\parallel}}{C}}-\underset{\underset{}{}}{\overset{\overset{H}{|}}{N}}-\underset{\underset{H}{|}}{\overset{\overset{\boxed{R'}}{|}}{C}}-COOH + H_2O$$

peptide bond

Figure 2.19

Peptide bonds between amino acids are formed by a condensation reaction. The dipeptide so formed still has one amino group terminal and one carboxyl group terminal that can make further peptide bonds with other amino acids.

in deoxyribonucleic acid (DNA). This determines the sequence of amino acids in a protein chain, as well as initiating and terminating the process. Changes in the body's structure and function are brought about by changing the extent to which the genetic potential is expressed.

Two amino acids form a dipeptide, and longer chains are known as polypeptides. Each polypeptide chain will have a free amino terminal and a free carboxyl terminal. Many proteins consist of more than one polypeptide chain, each of which forms a subunit. The order in which the amino acids occur is determined during protein synthesis by the sequence of the nucleotide bases in the particular DNA that contains the genetic information relating to that protein. The sequence of amino acids determines the ultimate structure as the side chains of the component amino acids attract, repel, or physically interfere with, each other, causing the molecule to fold and assume its final shape.

Protein structures are classified as primary, secondary, tertiary, and quaternary

The primary structure of proteins is defined as the linear sequence of amino acids joined together by peptide bonds. The position of covalent disulphide bonds between cysteine residues is also included in the primary structure. The secondary structure refers to the regular folding of regions of the polypeptide chain and results from short-range interactions between adjacent amino acid side groups. The two most common types of secondary structure are the α-helix and the β-pleated sheet (see Figure 2.20). The α-helix is a cylindrical, rod-like helical arrangement of amino acids in the polypeptide chain and is maintained by hydrogen bonds parallel to the helical axis. The rod-like structure of the myosin molecule's tail arises from the presence of a long polypeptide sequence in an α-helical configuration. In the β-pleated sheet, hydrogen bonds formed between adjacent sections of the polypeptide chain that are running in the same or opposite directions form a parallel or an antiparallel pleated sheet-like structure, respectively. The tertiary structure of the protein is the compact three-dimensional shape that results from non-covalent interaction of the side chains, and it confers catalytic as well as structural properties. The head region of the myosin molecule is globular in shape and has a catalytic region with ATPase activity (that is it can split ATP into ADP and P_i) as well as another region capable of binding to actin. Some proteins contain more than one subunit, giving a quaternary structure: haemoglobin, for example, consists of four separate polypeptide chains, two α-chains and two β-chains, as shown in Figure 2.21d. The overall structure allows changes in shape resulting from the binding of oxygen to alter the affinity of the other oxygen binding sites: in this way, the relationship between oxygen partial pressure and oxygen binding assumes a sigmoid rather than a linear shape, greatly increasing the sensitivity to small changes in the local partial pressure of oxygen. Actin monomers are globular in shape, but polymerize to form a chain called F-actin; this quaternary structure forms the thin filaments of the myofibrils in muscle.

Figure 2.20

The folding of the polypeptide chain into (a) an α-helix and (b) a β-pleated sheet. In the α-helix, the N–H of every peptide bond is hydrogen bonded to the C=O of a neighbouring peptide bond located four peptide bonds away in the same polypeptide chain. In the β-pleated sheet, adjacent peptide chains run in opposite directions and the chains are held together by hydrogen bonding between peptide bonds in different strands. The amino acid side chains in each strand alternately project above and below the plane of the sheet.

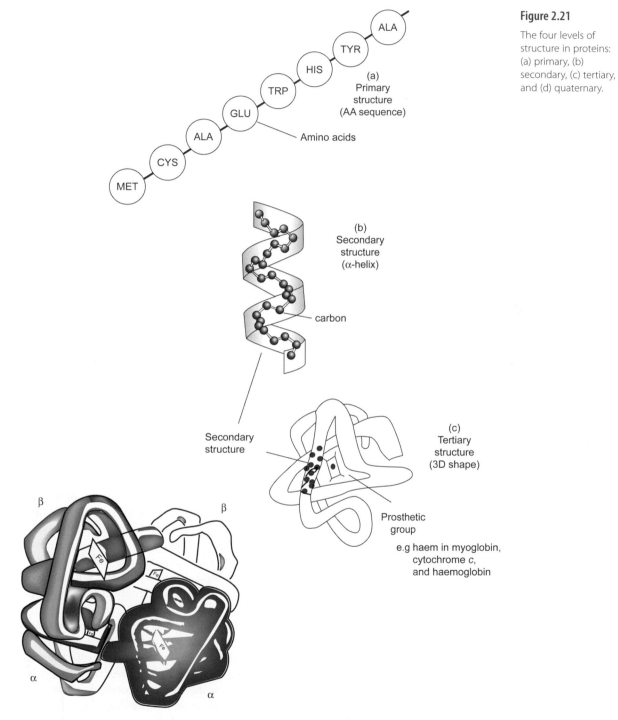

Figure 2.21

The four levels of structure in proteins: (a) primary, (b) secondary, (c) tertiary, and (d) quaternary.

(a) Primary structure (AA sequence)

Amino acids

ALA
TYR
HIS
TRP
GLU
ALA
CYS
MET

(b) Secondary structure (α-helix)

carbon

Secondary structure

(c) Tertiary structure (3D shape)

Prosthetic group

e.g haem in myoglobin, cytochrome *c*, and haemoglobin

β   β

α   α

(d) Quaternary structure (aggregation of subunits) e.g. haemoglobin

The activity of many enzymes is controlled by the reversible binding of regulatory agents, whether activators or inhibitors, that alter the shape of the active sites on the enzyme in such a way as to alter the affinity for the substrate. The tertiary and quaternary structures of proteins are sensitive to changes in pH and temperature. Most proteins undergo an irreversible change of three-dimensional structure when the temperature rises above about 45 °C. This process is called denaturation.

Protein turnover

Most structural proteins and enzymes are synthesized and degraded at high rates

There is no mechanism for storing excess dietary proteins in the body, and any amino acids that are ingested in excess of the immediate requirement are oxidized and the nitrogen excreted. Although the overall structure of the body is fairly stable, many of the component tissue proteins also have a relatively short life-span in the body. Most structural proteins and enzymes are synthesized and degraded at high rates, and as much as 20% of the basal rate of energy expenditure is the result of protein turnover. This process is obviously important in the repair of damaged tissue and in wound healing, but is also an ongoing process in healthy tissue. The half-life of some proteins is extremely short; for example, the half-life of some enzymes in liver is less than 1 h. Changes in the amount of these enzymes are an important factor in the control of their activity, and a high rate of turnover is therefore essential if the tissue is to be responsive to changes in metabolic requirements. In the liver, a high rate of turnover of the enzymes that regulate fuel homeostasis allows regulation to occur rapidly in response to feeding and to short-term fasting.

Some proteins are much more stable, with half-lives measured in days and weeks rather than hours: skeletal muscle adapts to training and detraining with a time course that can be measured in days, giving some indication of the rate of turnover of the enzymes involved in the process of adaptation. Breakdown of proteins into their component amino acids is achieved by hydrolytic enzymes (including a number of different proteinases and peptidases): these are derived from the lysosomes (small membranous vesicles containing various digestive enzymes), which can engulf and digest intracellular structures, but may also exist in soluble form. The mechanisms that control the activity of these enzymes are not well understood, but the process is clearly influenced by insulin, thyroid hormone, and several other factors. Protein turnover is a balance between rates of synthesis and degradation, and the concentration of these hormones, as well as of growth hormone, testosterone, cortisol, and other hormones, has effects on the rate of protein synthesis. The balance between the anabolic reactions that drive protein synthesis and the catabolic reactions that control protein degradation is disturbed by many factors, of which exercise is one. More dramatic effects are seen, however, in response to conditions such as infection, in which a high rate of protein degradation and nitrogen loss from the body occurs. An anabolic state is induced in skeletal muscle in response to weight training, but endurance training has no such effect. Although strength athletes have learned by experience the most effective ways to stimulate muscle hypertrophy, the mechanisms by which these effects are achieved are only just starting to be understood.

Protein breakdown can provide amino acids for various purposes

The breakdown of protein serves three main purposes:

1. It degrades potentially damaged proteins to prevent a decline in their function. Note that there is usually a net replacement of these degraded proteins.

2. It provides energy when some of its individual amino acids are converted into acetyl-CoA or TCA-cycle intermediates and are oxidized in the mitochondria.

3. Individual amino acids can be used for the synthesis of other compounds, including neurotransmitters (e.g. serotonin), hormones (e.g. adrenaline), and other peptides and proteins.

 This breakdown of protein and incorporation of the amino acids into a new protein links protein degradation with protein synthesis. Amino acids can also be incorporated into compounds that are not proteins.. For example, the carbon skeletons of some amino acids are converted into glucose (gluconeogenesis), ketones (ketogenesis), or fat (lipogenesis) and subsequently stored in adipose tissue. Some amino acids can also be oxidized and so can serve as a fuel for muscle, particularly during prolonged exercise. Most people eat far more protein that they need to for the synthesis of new proteins and other amino acid derivatives, and there is no way of storing this excess in the form of protein. Oxidation is the most effective way of disposing of the excess, but this does increase the amount of nitrogen and sulphur that must be excreted. These aspects of amino acid metabolism are covered in more detail in Chapter 5.

Exercise affects rates of protein turnover and both rates of protein synthesis and breakdown are altered

After resistance exercise, muscle protein turnover is increased because of an acceleration of both protein synthesis and degradation. Muscle protein breakdown is increased after resistance exercise but to a smaller degree than muscle protein synthesis. The elevations in protein degradation and synthesis are transient but are still present at 3 h and 24 h after exercise, although protein turnover returns to baseline levels after 48 h. These results seem to apply to resistance exercise or dynamic exercise at a relatively high intensity. Low-intensity to moderate-intensity dynamic endurance exercise does not seem to have the same effects on muscle protein turnover, although studies have shown that endurance exercise may result in increased protein oxidation, especially during the later stages of very prolonged exercise and in conditions of glycogen depletion. Interestingly, studies have shown that the body adapts to training by becoming more efficient with protein. Protein turnover decreases after training, and less net protein degradation occurs. In other words, after training, athletes become more efficient and 'waste' less protein.

Anabolic hormones

 Some hormones have anabolic effects

In the normal adult human, a number of hormones are known to have anabolic effects. Growth hormone is secreted by the pituitary gland under the control of the hypothalamus. Large increases in growth hormone output—or at least in its release into the circulation—have been reported to occur after short-term, high-intensity exercise, but the significance is unclear. Growth hormone is known to antagonize some of the actions of insulin, and also acts on the liver to cause the release of a family of peptides known as somatomedins. These include the insulin-like growth factors (IGF-1 and IGF-2): these peptides have a general anabolic effect similar to that of insulin. It is known that athletes in strength events as well as bodybuilders have been using injections of human growth hormone, often in combination with insulin, in an attempt to stimulate muscle protein synthesis and hence improve performance. These attempts have been only partially successful as growth hormone is more effective in stimulating the synthesis of collagen, which forms the connective tissue in muscle, rather than promoting increases in the synthesis of the contractile proteins actin and myosin. Improvements in performance are therefore somewhat less than generally anticipated. There are also negative consequences resulting from the administration of growth hormone, including growth of some of the bones of the face, hands and feet, as well as increasing the predisposition to diabetes.

Insulin is well known to stimulate protein synthesis, but the mechanism by which this is achieved and the specificity of the effect are unclear. Testosterone has an anabolic effect that is important during the adolescent growth spurt, and abuse of testosterone and related anabolic steroids by athletes indicates the extent to which muscle growth can be stimulated by a combination of training and steroid administration. Neither training nor steroid administration alone appears capable of achieving the degree of hypertrophy that many champion bodybuilders demonstrate. Again, however, chronic use of steroids is not without major health risks, and deaths from cardiovascular disease at a relatively young age, as well as from cancers of the liver, are relatively common in those who have used these drugs.

	Synthesis	Degradation
Physical activity		Decrease
Leucine	Increase	
Insulin	Increase	Decrease
Glucagon		Increase
Glucose		Decrease
Testosterone	Increase	
Glucocorticoids		Increase
Triiodothyronine (T3)	Increase	

Table 2.5 Some of the factors that are known to influence protein synthesis and degradation in skeletal muscle.

   As might be expected from the known antagonism of the actions of glucagon and insulin, an increase in glucagon activity stimulates protein breakdown. High levels of the thyroid hormone triiodothyronine also increase the rate of protein degradation, although low levels also stimulate synthesis. The mechanism of action of these hormones is again unclear. Muscle activity has a powerful effect on the rate of protein degradation, as is readily seen if a muscle is immobilized: within a few days, a dramatic loss of muscle mass occurs. A summary of some of the factors that influence protein synthesis and degradation is given in Table 2.5.

Proteins as enzymes

Enzymes speed up the rate of specific chemical reactions

Enzymes are proteins that act as controllable catalysts: they speed up the rate of specific chemical reactions and allow them to be regulated in a way that permits the body to control the interactions between the different metabolic pathways that sustain life. The direction in which the reactions proceed and the equilibrium point that would be reached in a non-biological system are governed by the laws of thermodynamics. The most striking feature of enzyme-catalysed reactions is that they can reach the point of saturation. The characteristics of enzymes that cause this behaviour are described briefly here.

Mechanisms of enzyme action and enzyme kinetics

The formation of an enzyme substrate complex lowers the energy of activation

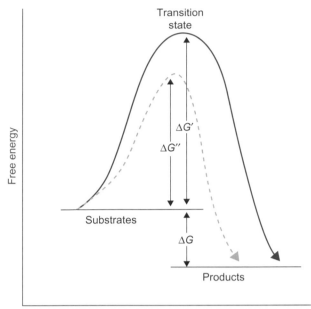

Figure 2.22

The total free energy of the reaction products is less (by an amount equal to ΔG) than that of the substrates. In order to allow the reaction to proceed, however, sufficient energy to overcome the energy of activation ($\Delta G'$) must be available. An enzyme lowers the energy of activation ($\Delta G''$), making it more likely that the reaction will proceed.

The laws of thermodynamics tell us that chemical reactions proceed spontaneously only in the direction that results in the products of the reaction having a lower energy status than the substrates (Figure 2.22). Enzymes function by acting as reusable catalysts: this involves the formation of an enzyme substrate complex as an intermediate step in the reaction. The formation of this intermediate step lowers the energy of activation. Because less energy now has to be added, the reaction is more likely to proceed. It is important to appreciate that the enzyme, although it participates in the reaction, is not consumed and is therefore required to be present in only small amounts.

The energetics of formation of the enzyme substrate complex are not well understood, but there is clearly some kind of weak bond formed between the substrate and the enzyme. This involves one or more active sites on the enzyme, and these sites have a particular shape and charge distribution that allows them to interact with the substrate. These characteristics allow enzymes to promote the rates of specific reaction in a number of different ways. Where two or more substrates are involved, attachment to binding sites on the enzyme allows the substrates to be brought into close proximity in the correct orientation, thus increasing the chances of a reaction taking place. Alternatively, binding to the enzyme can cause changes in the shape of the substrate molecule that increase its susceptibility to reaction.

Enzyme kinetics refers to the measurement of the change in substrate or product concentration as a function of time. The first stage in an enzyme-catalysed reaction is the binding of the substrate (S) to the active site of the enzyme (E) to form an enzyme–substrate complex (ES): the substrate then reacts to form the product (P), which is released. Release of the product restores the enzyme to its original free form:

$$E + S \leftrightarrow ES \rightarrow E + P.$$

The assumption made here is that the first stage of the process is reversible, but that the second is not. In almost all reactions, the concentration of the substrate is far in excess of the enzyme concentration. This means that formation of the ES complex does not result in an appreciable change in the substrate concentration, but does reduce the concentration of the free enzyme.

You can see from Figure 2.23 that the progress curve for the reaction is initially linear, decreasing in slope as the reaction proceeds and substrate is used up. The initial velocity during the linear part of the curve is called V_0. The relationship between V_0 and the substrate concentration ([S]) is described by the Michaelis–Menten equation:

$$V_0 = V_{max} [S]/K_m + [S],$$

where V_{max} is the maximum velocity of the reaction at infinite [S] and K_m is the Michaelis constant, equivalent to the substrate concentration where the initial reaction velocity (V_0) is equal to half the maximal velocity (i.e. $K_m = [S]$, where $V_0 = V_{max}/2$). This relationship is depicted graphically in Figure 2.23 and clearly shows that at low concentrations of substrate, the initial reaction rate increases linearly in response to increasing substrate concentration, but the rate approaches a limit above which it is constant and independent of substrate concentration. At this point, all of the enzyme molecules are effectively saturated with substrate. The V_{max} is therefore a function of the amount of enzyme present.

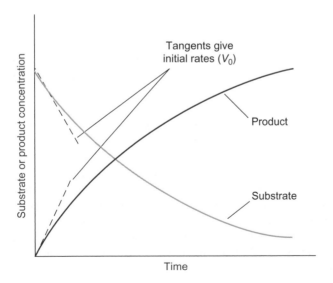

Figure 2.23

The progress curve of an enzyme-catalysed reaction: substrate utilization and product formation as a function of time. The tangents to the curves give the initial reaction rates (or initial velocities, V_0).

It also follows that when the substrate concentration is equal to K_m, the reaction rate will be equal to half of the V_{max}. The K_m value is therefore equal to the substrate concentration that will result in the reaction proceeding at one half of the maximum rate. A high K_m value is therefore an indication of low affinity of the enzyme for its substrate: a high substrate concentration is necessary to achieve a reaction rate equal to half of the maximum rate.

High reaction rates will be achieved only when the substrate concentration is relatively high. If [S] is equal to 10 times the K_m, substituting these values into the Michaelis–Menten equation tells us that the reaction rate will be 91% of V_{max}, while 99% of the maximum rate will only be achieved when the substrate concentration is 100 times the K_m. For a detailed description of the derivation of the Michaelis–Menten equation see Cornish-Bowden and Wharton (1988).

Enzyme activity can be measured

The activity of enzymes can be assessed by the rate of substrate utilization or product formation under standardized conditions. The most common unit of measurement is the international unit (U or IU). This is the amount of enzyme that will convert 1 μmole of substrate to product in 1 min under the conditions that are specified for that reaction. Although this is the generally used measure among physiologists, the appropriate SI unit should be used. This is the katal (kat), which is defined as the amount of enzyme that will convert 1 mole of substrate to product in 1 s under optimum conditions. At least part of the reason for the persistence of the IU is the difficulty in defining optimum conditions for the activity of individual enzymes.

Several factors, including temperature and pH, influence enzyme activity

Enzyme activity is particularly sensitive to temperature, and will increase as the temperature increases. Any expression of enzyme activities must therefore specify the temperature at which measurements are made: temperatures of 25 and 37 °C are normally used as standards. At high temperatures, however, enzyme activity falls sharply and irreversibly because of structural changes caused by denaturation of the protein (see Figure 2.25a). Although body-core temperature is usually about 37 °C, the temperature of muscle tissue may be as low as 30 °C in a resting person on a cold day and can rise to 42 °C during high-intensity exercise. Hence, warming up before an event has important implications for maximizing reaction rates and optimizing muscle performance. Except in extreme cases of heat illness, body-core temperature seldom exceeds 41 °C, but this is close to the level at which some enzymes and other proteins are affected.

Changes in the ionization state of an enzyme caused by a change in pH of the cell will affect its affinity for its substrate because of changes in structure or charge distribution at the active site (Figure 2.26). The local pH may also affect

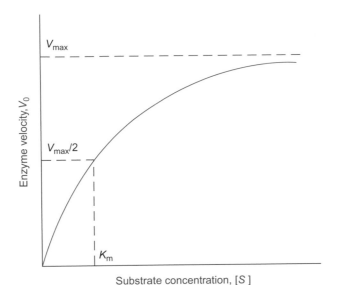

Figure 2.24

The effect of substrate concentration on enzyme activity. V_{max} is the maximum velocity of the reaction and depends on the amount of enzyme present. The substrate concentration at half the maximum velocity ($V_{max}/2$) is the Michaelis constant V_m.

the ionization state of the substrate. All enzymes have an optimum pH (as illustrated in Figure 2.25b), but this differs between different enzymes, and may also be influenced by the presence of other activators and inhibitors. Variations in pH are generally small in most tissues, with skeletal muscle showing the largest changes in response to very high intensity exercise: pH may fall from the resting value of about 7.1 to 6.5 or even less. Many enzymes normally function in an environment that is close to their pH optimum; for example pepsin, which has a pH optimum of about 2.0, seems well adapted for the acid conditions of the stomach where it acts to hydrolyse proteins into smaller fragments (peptides) and amino acids that can be absorbed in the small intestine. Some enzymes, however, have a pH optimum, at least in their isolated and purified form, that is far from their normal environment; glycerol kinase demonstrates maximum activity at a pH of 9.8, a condition that is never reached in the cell.

Many enzymes require coenzymes, prosthetic groups, or cofactors

Many enzymes require the presence of one or more coenzymes if the reaction is to proceed: these are compounds that participate in the reaction. For example, the conversion of lactate to pyruvate, which involves the removal of two hydrogen atoms from lactate and is catalysed by lactate dehydrogenase, requires that the coenzyme nicotinamide adenine dinucleotide (NAD^+) be available to participate in the reaction. Coenzymes are chemically altered by participation in the reaction, in this case by conversion of NAD^+ to its reduced form, NADH. The coenzyme is therefore essentially a substrate for or product of the reaction,

Figure 2.25

The effect of (a) temperature and (b) pH on enzyme activity.

(a)

(b)

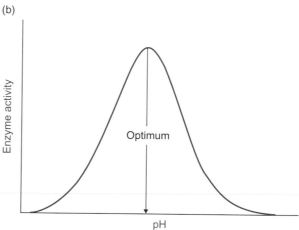

but a characteristic of coenzymes is that they are readily regenerated by other reactions within the cell. Some coenzymes, such as NAD^+, are loosely bound to the enzyme, but others (e.g. biotin) are tightly bound and are referred to as prosthetic groups.

Many enzymes have only low activities in the absence of cofactors, and the presence of a metal ion, especially the divalent metals calcium, magnesium, manganese, and zinc, is essential for activation of many enzymes. Binding of these ions alters the charge distribution and shape of the active site of the enzyme. The release of calcium into the cytoplasm in response to the nerve impulse is important in the activation of phosphorylase, which allows acceleration of the glycolytic pathway.

Competitive and non-competitive inhibition may interfere with enzyme function

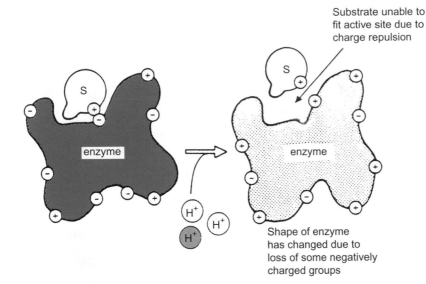

Substrate unable to
fit active site due to
charge repulsion

Shape of enzyme
has changed due to
loss of some negatively
charged groups

Figure 2.26

Addition of protons (i.e. a fall in pH) will remove some negatively charged groups and make additional positively charged groups on the protein molecule. This may hinder or prevent the substrate from binding to the active site of the enzyme. Raising the pH will reverse these changes.

Substances that have a chemical structure similar to that of the normal substrate for an enzyme may also be able to bind to the active site on the enzyme and thus interfere with enzyme function by reducing the number of active sites that are available to the proper substrate. These substances compete with the substrate for access to the active site and are therefore referred to as competitive inhibitors. The effect of competitive inhibition is to increase K_m. Increasing the concentration of substrate to a sufficient level will, however, swamp the effects of the inhibitor and V_{max} is not affected by competitive inhibition.

Non-competitive inhibitors bind to the enzyme at other sites, leaving the active site of the enzyme available to the substrate, but they have the effect of altering the conformation of the protein and thus reducing the catalytic activity of the active site. The V_{max} is reduced, but the same substrate concentration will still produce one half of the new maximum activity; that is K_m remains unchanged.

Control of enzyme activity

Allosteric and covalent modulation can be used to regulate enzyme activity

Allosteric modulation of enzyme activity refers to the reversible binding of small molecules to the enzyme at sites other than the active site, producing a conformational change in the structure of the enzyme molecule. This change in shape and charge distribution results in a change (either an increase or a decrease) in the affinity of the enzyme for its substrates or products and hence

Figure 2.27

(a) Allosteric modulation of enzyme activity. Binding of a specific modulator molecule at a site other than the functional site of the enzyme causes a shape change in the enzyme molecule that results, in the case shown here, in the activation of the functional (active) site of the enzyme. Note that in other cases, allosteric modulators can also cause inhibition of enzyme activity. (b) Covalent modulation of enzyme activity. Addition of an inorganic phosphate (P_i) group at a site other than the functional site of the enzyme causes a shape change in the enzyme molecule that causes, in the case shown here, the activation of the functional (active) site of the enzyme. Note that in other cases, covalent modulation can also cause inhibition of enzyme activity.

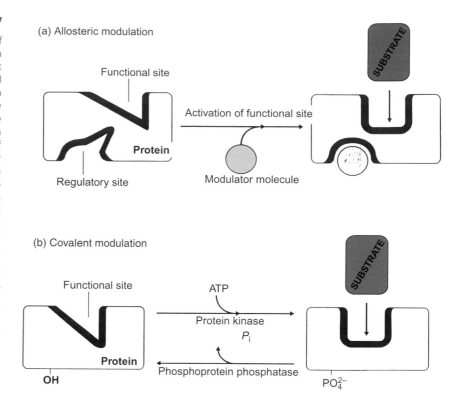

(a) Allosteric modulation

Functional site

Protein

Regulatory site

Activation of functional site

Modulator molecule

SUBSTRATE

(b) Covalent modulation

Functional site

Protein

OH

ATP

Protein kinase

P_i

Phosphoprotein phosphatase

PO_4^{2-}

SUBSTRATE

in its activity (Figure 2.27a). As you can see in Figure 2.27b, covalent modulation involves phosphorylation or dephosphorylation of an enzyme—usually involving the hydroxyl ($-OH$) group of a serine residue in the polypeptide chain. The importance of covalent modulation in controlling the activity of some of the key enzymes of glycolysis is described in Chapter 4.

Enzyme isoforms

Many enzymes exist in more than one form: these are called isoforms

Many enzymes exist in more than one form: these isoforms catalyse the same reaction, but are generally found in different tissues and may have different specificities or catalytic capabilities. Lactate dehydrogenase exists in two forms, each made up of four subunits. The subunits exist in one of two forms: the H form is found predominantly in cardiac muscle, whereas the M form predominates in skeletal muscle. Five different combinations of these subunits are possible. In muscle, the H form is associated with tissues that have a high capacity for oxidative metabolism, and therefore have a high capacity for lactate oxidation, whereas the M form is associated with tissues having a high anaerobic capacity relative to their oxidative capability. The M form favours the conversion of pyruvate to lactate, whereas the H form favours the conversion of lactate to pyruvate. Many other enzymes also exist in a variety of different isoforms, but the functional significance of these is not well understood.

Energy for muscle contraction

The role of ATP

Force generation by skeletal muscles requires a source of chemical energy in the form of ATP

Energy can be defined as the potential for performing work or producing force. Development of force by skeletal muscles requires a source of chemical energy in the form of ATP; in fact, energy from the hydrolysis of ATP is harnessed to power all forms of biological work. In skeletal muscle, energy from the hydrolysis of ATP by myosin ATPase activates specific sites on the contractile elements, as described previously, causing the muscle fibre to shorten. Active reuptake of calcium ions by the sarcoplasmic reticulum also requires ATP, as does the sodium/potassium pump in the sarcolemmal membrane that helps to restore the muscle cell's resting membrane potential (Figure 2.28). There are four different mechanisms involved in the generation of energy for muscle contraction:

1. ATP is broken down enzymatically to adenosine diphosphate (ADP) and inorganic phosphate (P_i) to yield energy for muscle contraction.
2. Phosphocreatine (PCr) is broken down enzymatically to creatine and phosphate, which are transferred to ADP to re-form ATP.
3. Glucose 6-phosphate, derived from muscle glycogen or blood-borne glucose through anaerobic glycolysis, is converted to lactate and produces ATP by substrate-level phosphorylation reactions.
4. The products of carbohydrate, lipid, protein, and alcohol metabolism can enter the tricarboxylic acid (TCA or Krebs) cycle in the mitochondria and be oxidized to carbon dioxide and water. This process is known as oxidative phosphorylation and yields energy for the synthesis of ATP.

The utilization of ATP as the immediate source of energy is the first of these mechanisms and the purpose of the other three mechanisms is to regenerate ATP at sufficient rates to prevent a significant fall in the intramuscular ATP concentration. The reversal of reactions 1, 2, and 3 above ultimately depends on energy released by oxidative metabolism.

Figure 2.28

The main ATP-requiring processes during muscle contraction and the principal means of regenerating ATP.

 Very rapid powerful contractions rely heavily on available ATP

For the weightlifter, we can limit our consideration to the role of ATP. As the lifting of heavy weights occurs within a matter of only a second or two, it is only the energy immediately available to the muscle in the form of ATP that can be used. The other means of resynthesizing ATP take too long to make any substantial contribution during the power lift of the event itself. The hydrolysis of a molecule of ATP to ADP and P_i yields a certain amount of *free energy*. According to the second law of thermodynamics, during any biochemical reaction a certain amount of energy is transformed into a random disordered form that is unavailable to do work (*entropy*). In addition, *free energy* and *heat energy* (*enthalpy*) also become available. The former, which is usually signified by the symbol G, can be used to do work at a constant temperature and pressure. It is normally the case that energy transformation in skeletal muscle takes place at a constant temperature and pressure, and therefore the changes in free energy, enthalpy, and entropy can be described by the following equation:

$$\Delta G = \Delta H - T\Delta S,$$

where ΔG is the change in free energy, ΔH is the change in enthalpy, T is the absolute temperature (in kelvin, K), and ΔS is the change in entropy. Note that it is the changes in free energy, enthalpy, and entropy that are used rather than their absolute values.

Every reaction has a characteristic ΔG. Furthermore, it can be seen from the above equation that it is possible to calculate the free energy of any biochemical reaction, assuming standard conditions of temperature (310 K), pressure

(101 kPa), and pH (7.0). The simplest definition of ΔG for a biochemical reaction is the difference in free-energy content between the reactants and the products under standard conditions. For example, assuming the following reaction:

$$A + B \leftrightarrow C + D,$$

ΔG is less than zero when the sum of the free-energy contents of C and D is less than that of A and B. In this situation, the reaction tends to occur spontaneously in a rightward direction. If ΔG is positive the reaction cannot occur spontaneously in this direction and will only occur if energy is added to the system, but the reaction will occur spontaneously in the opposite direction without the addition of energy. If the ΔG is zero the reaction proceeds in neither direction and is said to be in a state of equilibrium.

Under standard conditions, the standard reaction free energy is obtained:

$$\Delta G^\circ = \Sigma \Delta G^\circ \text{ products} - \Sigma \Delta G^\circ \text{ substrates} = \Delta H^\circ - T\Delta S^\circ.$$

The *standard free energy of formation,* ΔG°_f, is the standard free energy of the formation of a mole of substance from its elements under a standard set of conditions.

The actual amount of free energy released in a reaction under given conditions depends on the concentrations of the substrates and products of the reaction and can be calculated if the concentrations of the substrates and products are known or estimated. The actual free energy of the reaction is the difference between the free energies of the products and the free energies of the reactants and the free energies of formation of both the substrates and products must be multiplied by the numbers of moles present. The standard free energy for the splitting of ATP to the products ADP and P_i is -31 kJ/mol and the actual free energy of ATP hydrolysis (ΔG_{ATP}) is determined as:

$$\Delta G_{ATP} = \Delta G^\circ_{ATP} + RT \ln ([ADP] \times [P_i] / [ATP]),$$

where R, the gas constant, is 8.34 kJ/mol K $\times 10^{-3}$ and T, the absolute temperature is 310 K.

Based on the actual measured concentrations of ATP, ADP, and P_i in resting human skeletal muscle, the actual amount of free energy from the splitting of ATP is more negative than -31 kJ/mol and has been estimated to be between -58 and -63 kJ/mol. To illustrate the importance of taking substrate and product concentrations into account when estimating the amount of free energy available, consider putting very high ADP and P_i concentrations in a test tube and very little ATP. Under these conditions the reaction would reverse (ADP and P_i would spontaneously form ATP) or energy would be needed to split ATP even further into ADP and P_i. Thus, the closer we get to chemical equilibrium, the less energy we gain from the reaction proceeding.

Burning fuel to resynthesize ATP

Free energy released during breakdown of carbohydrates and lipids can be
stored in the ATP molecule

Free energy released during combustion of carbohydrates and lipids can be stored
in the compound ATP, hence the terms 'high-energy phosphate' or 'phosphagen',
which are commonly used to describe this compound. ATP is the only form
of chemical energy that can be converted into other forms of energy used by
living cells. ATP can be used as the source of energy for biosynthesis of macromolecules, for membrane transport against prevailing concentration gradients,
and for mechanical work including muscle contraction. Reactions involving the
breakdown of the phosphate bonds and the liberation of P_i are catalysed by
enzymes called kinases. In the case of ATP breakdown, these are commonly
abbreviated to ATPases. Furthermore, the cellular concentrations of ATP, ADP,
and P_i are such that the ΔG of the adenylate kinase reactions (also known as
the myokinase reactions) depicted below are large and negative (approximately
$58-63$ kJ per mole of ATP), thereby enabling enough chemical energy to be
converted into mechanical work to produce force or movement.

$$\text{ATP} \rightarrow \text{ADP} + P_i \text{ and } \text{ATP} \rightarrow \text{AMP} + 2P_i.$$

Muscles are able to perform up to 24 kJ of work for each mole of ATP degraded.
It follows, therefore, that the ΔG for ATP hydrolysis must be greater than 24 kJ/mol.
It is also worth noting that the component parts of the adenylate kinase reaction
are interconvertible without any net change in ΔG:

$$\text{ATP} + \text{AMP} \rightarrow \text{ADP} + \text{ADP}.$$

Look at the molecular structure of ATP in Figure 2.29a. As you can see, it is a
nucleotide consisting of the purine base adenine, the five-carbon sugar ribose
and a triphosphate unit. Generally, the most metabolically active form of ATP is
the magnesium salt and the most interesting part of the compound in terms of
exercise metabolism is the triphosphate unit. The sequential hydrolysis of the two
terminal phosphate bonds in the adenylate kinase reactions releases a substantial
amount of free energy that is used to drive the numerous energy-requiring reactions and processes in the cell:

$$\text{ATP} + \text{H}_2\text{O} \rightarrow \text{ADP} + P_i + \text{H}^+$$
$$\text{ADP} + \text{H}_2\text{O} \rightarrow \text{AMP} + P_i + \text{H}^+.$$

It should be noted that there is nothing particularly special about the phosphate
bonds in the above reactions. Their importance lies in the fact that they release
free energy when hydrolysed. Note that the amount of energy released is more
than the maximum amount of work that can be done (24 kJ/mol ATP), which

Figure 2.29

(a) The structure of adenosine triphosphate (ATP). Note that ATP consists of an adenine group (purine), a ribose group (sugar), and a triphosphate unit. ATP is broken down by the action of ATPase enzymes. The addition of water (hydrolysis) releases ADP, inorganic phosphate, a hydrogen ion, and energy.

reflects inefficiency in the transformation of the chemical energy to mechanical energy in the contractile mechanism. Also note that the breakdown of ATP by hydrolysis generates P_i and H^+ ions (protons) as shown in Figure 2.29b. In sustained contractions the accumulation of these products of ATP breakdown contributes to fatigue.

ATP is the energy currency of the cell

Because of its unique role in energy production, ATP has been termed the energy currency of the cell. However, unlike money, ATP cannot be accumulated in large amounts and the intramuscular ATP store is limited to 5 mmol/kg wet weight (ww) of muscle. Indeed, given that changes in the cellular concentrations of ATP, ADP, and AMP are involved in the regulation of many metabolic processes, it would be unwise to simply have a larger cellular store of ATP. During maximal exercise, there is sufficient ATP present to fuel about 2 s of contraction. However, the muscle ATP store never becomes completely depleted because it is normally efficiently resynthesized from ADP and AMP at the same rate at which it is degraded. During submaximal, steady-state exercise this is achieved by mitochondrial oxidation of carbohydrate and fat. However, during the rest to steady-state transition period at the onset of submaximal exercise and during high-intensity exercise lasting more than a few seconds this is achieved principally by anaerobic ATP resynthesis from PCr breakdown and glycolysis. Further details of these important pathways are given in Chapters 3 and 4.

Nutritional effects on strength training and performance

Protein needs

Weightlifters do not need very large intakes of protein

Athletes taking part in strength and power events are recognized for their concern with diet, and there is some justification for this as the foods that are eaten will affect performance. Because muscles are made of protein, it is tempting to believe that increasing the amount of protein in the diet will help to build muscle: a high-protein diet is indeed a characteristic of these athletes. The available evidence does suggest an increased protein need in strength athletes. The average individual needs about 0.6 g of protein per kilogram of body mass per day to stay healthy, and the recommendation is that about 0.8 g/kg/day should be consumed by the average individual. This provides a bit extra to allow for individual variation in requirements and also allows for differences in the amino acid content of the different proteins that are found in the diet. With increasing food intake the intake of protein automatically increases because many food products contain at least some protein. There is usually a linear relationship between energy intake and protein intake. It is often recommended that strength athletes should eat about 1.4–1.7 g/kg/day, although intakes well in excess of 2–3 g/kg/day are not unusual.

Although the scientists say that such high intakes are not necessary, weightlifters and bodybuilders believe otherwise. The reason for this divergence between scientific evidence and popular belief may lie in the fact that most scientific studies have not measured elite athletes and have looked at only short time periods. One potential problem with the habitual consumption of a high-protein diet is that the enzymes involved in protein breakdown may be upregulated. This, in effect, means that the athlete becomes dependent on a high-protein diet to maintain muscle mass as any reduction in intake means that breakdown continues at a high rate while the rate of protein synthesis will be reduced because of the reduced availability of amino acids from the diet. The weightlifter or bodybuilder who suddenly reduces protein intake to the levels recommended by the scientists will therefore experience a loss of muscle. Excessive protein intake [more than 3 g/kg bodyweight (bw)/day] may have various negative effects including kidney damage, increased blood lipoprotein levels (which has been associated with the development of arteriosclerosis), and dehydration.

> The timing of food intake in relation to exercise may be important to maximize adaptation to training

After each resistance training session, the rates of protein synthesis and protein breakdown in the exercised muscles are both increased. Training in the fasted state limits the gains in muscle mass that can be made. Increasing the rate of synthesis means that the free amino acid pool in the muscle falls and this limits the rate of synthesis. If additional amino acids are supplied by the diet at this time, protein synthesis is increased. There are now several laboratory studies showing that ingestion of protein (typically about 40 g) or mixtures of the essential amino acids (typically about 6 g) results in increases in net protein synthesis compared with control trials where the subjects remain fasted. The major stimulator of muscle protein synthesis is the blood amino acid concentration (which increases 2–3-fold after a meal containing protein). It is the essential amino acid component that is stimulatory, especially the branched-chain amino acids (BCAA) and particularly leucine. However, all 20 amino acids are required for protein synthesis to take place. The stimulatory effect of essential amino acids on protein synthesis (approx doubling of the synthetic rate) occurs by three mechanisms:

(1) The mass action effect (i.e. more substrate available for protein synthesis).
(2) A leucine sensor on the outside of cells (it must be on the outside because when essential amino acids are infused into humans raising the blood amino acid concentration, the intracellular amino

acid concentration decreases—suggesting that amino acids inside the cell are being channelled into protein synthesis) directly stimulates protein synthesis. Leucine appears to activate the phosphorylation of mTOR, a regulatory protein that activates the anabolic signalling pathway.

(3) Rise in essential amino acid concentrations in blood stimulates insulin secretion that stimulates muscle uptake of amino acids and (in animals, less so in humans) it stimulates protein synthesis.

Larger rises in circulating insulin can be achieved with the coingestion of carbohydrate than with protein ingestion alone and recent studies have confirmed that increases in net tissue protein gain are highest when a combination of carbohydrate and protein is ingested shortly after a resistance exercise training session. The stimulation of protein synthesis with essential amino acids seems less effective in the elderly, which may account for muscle wasting with ageing. The amounts of anabolic signalling proteins in the elderly are diminished.

Supplements

Many dietary supplements that are claimed to increase muscle mass and strength are ineffective

Supplementation of the diet with individual amino acids is also popular, even though there is little evidence that this is effective in building muscle. Arginine and ornithine, in high doses, can stimulate release of growth hormone, but strength training itself has this effect. Normal protein-containing foods supply all of the essential amino acids, so the reasons for using these supplements are not clear. Branched-chain amino acids are among the most popular nutrition supplements. However, the evidence for claims of reduced net protein breakdown and reduced fatigue and enhanced performance via central nervous mechanisms are not convincing.

Strength and power athletes use many different dietary supplements, but most of these are almost certainly completely ineffective. Beta-hydroxy beta-methylbutyrate (HMB) is a metabolite of the essential amino acid leucine. Its use as a supplement has increased dramatically in the past few years especially among bodybuilders. It is claimed that HMB increases lean body mass and strength and improves recovery from heavy exercise. HMB is synthesized in the human body at an estimated rate of about 0.2–0.4 g/day.

$$\text{Leucine} \xrightarrow{\text{(Transaminase)}} \alpha\text{-keto isocaproate (KIC)} \xrightarrow{\text{(KIC-dioxygenase)}} \text{HMB.}$$

Studies in rats have demonstrated that supplementing the amino acid leucine can be anticatabolic. Studies have also investigated HMB as a potential anticatabolic agent in farm animals. It has been hypothesized that HMB could reduce protein breakdown in humans, resulting in increased muscle mass and strength. HMB supplementation has been found to decrease protein breakdown and slightly increase protein synthesis in rats and chicks, but has no effect on protein metabolism in growing lambs. Information about the effects of HMB in humans is scarce. The only published study currently available (Nissen *et al.* 1996a) studied 41 male untrained volunteers who participated in a resistance training programme for three weeks. The programme consisted of three 90-min weightlifting sessions per week. Participants were divided into three groups and each group received a different dose of HMB. The first group received placebo, the second group received 1.5 g of HMB per day, and the third group 3.0 g of HMB per day. Lean tissue tended to increase more in the HMB groups and this occurred in a dose-dependent manner. A point of criticism can be raised, however, because the group that had the largest increase in lean body mass (receiving 3 g HMB/day) was also the group with the lowest lean body mass and muscle strength to begin with. Therefore, this group could be expected to gain more than the placebo group, who already had a larger lean body mass. This study also had no diet control, so the dietary leucine intake was not known.

In part 2 of the same study (Nissen *et al.* 1996b), subjects trained for 7 weeks, 6 days per week, and they were supplemented with 3.0 g/day of HMB or placebo. This study showed an increase in fat-free mass in the HMB group after 14 days. Interestingly, after 39 days no differences were observed. Strength was also measured in this second study but no differences were found in the strength measurements except for the performance in a bench press. There was a small improvement in bench press strength with HMB supplementation (2.6 kg improvement after 7 weeks of training with HMB versus 1.1 kg improvement with placebo). A point of criticism again is the fact that diet was not controlled.

Although these studies suggest that HMB may have an effect on lean body mass and strength, the evidence is not convincing and more studies need to be conducted. It is also important to keep in mind that commercially available 'recovery' products often contain extremely small amounts of HMB while the above studies suggest that 3 g of HMB should be ingested to exert the desired effects.

Supplementing the diet with creatine seems to increase muscle strength and power

Supplementing the diet with creatine does seem to be effective in increasing strength and improving power. Most of the body's creatine (about 95%) is present in skeletal muscle and about two-thirds of this is in the form of phosphocreatine. Meat in the diet typically provides about 1 g/day of creatine, but vegetarians have almost no intake from the diet. The liver and kidney can synthesize creatine from the amino acids methionine, arginine, and glycine. Creatine is broken down to creatinine, which is excreted in the urine and about 2 g/day is lost by this process. Supplementing the diet with 10–20 g/day of creatine for about

4–6 days can increase the creatine content of muscle, improve strength and power, and typically results in an increase in body mass of 1–2 kg. This weight gain is lean tissue (mostly water), not fat.

Steroids

Anabolic steroids mimic the actions of testosterone

A range of drugs are also used by some athletes in the strength and power events: these include especially anabolic steroids, which mimic the actions of testosterone in promoting both muscle growth and aggression. Androstenedione is a potent anabolic steroid and is one of the most popular nutrition supplements in the USA. It is believed to be a precursor of testosterone in the body. By elevating testosterone synthesis it is believed to increase protein synthesis, build muscle mass, and improve recovery from exercise. Androstenedione was first developed in East Germany to enhance the performance of their athletes. A few years ago it became widely available on the market in the USA. The regulations in the Dietary Supplement Health and Education Act (1994) made it possible to sell androstenedione as a food supplement. It is now available over the counter in almost any drug store or pharmacy in the USA and can also be purchased from suppliers via the internet.

The evidence that androstenedione has anabolic properties, however, is far from convincing. Although only a very small number of studies have investigated the effects of androstenedione on serum testosterone concentrations and increases in muscle strength, we can draw some conclusions from these studies. One study by King and colleagues (1999) determined the effects of short-term and long-term (8 weeks) oral androstenedione supplementation (300 mg/day) on serum testosterone and oestrogen concentrations and skeletal muscle fibre size and strength in a group of individuals following a strength training programme. The group of 20 was divided randomly into a placebo group and an androstenedione group. Interestingly, no changes were observed in serum testosterone concentrations but serum oestradiol concentrations were increased after androstenedione administration. Strength training resulted in increased strength, increased lean body mass and increased cross-sectional area of Type-II (fast-twitch) muscle fibres after 8 weeks but there were no differences between the androstenedione and the placebo groups. A lack of effect on serum testosterone concentrations was also reported by three other studies in which 100–200 mg of androstenedione were ingested daily for 2 days to 12 weeks. One recent study by Rasmussen and co-workers (2000) did not find any effects on protein synthesis and breakdown or phenylalanine balance across the leg. Therefore, it seems reasonable to conclude that androstenedione has no effect on the plasma testosterone concentration, does not change protein metabolism, has no anabolic effect, and does not alter the adaptations to resistance training.

Androstenedione may have some negative health effects. Most anabolic steroids are harmful and produce a variety of side-effects including acne, facial and body hair growth, growth of the prostate, impaired testicular function, and liver damage. In fact, in one study a decrease in serum high-density lipoprotein (HDL)-cholesterol was observed, which has been associated with increased risk of cardiovascular disease.

Other drugs include stimulants and β-agonists, which simulate the effects of the catecholamine hormones, preparing the body for fight or flight. Of course, there are also ethical issues involved. Anabolic steroids including androstenedione and stimulants such as ephedrine and pseudoephedrine are banned by the World Anti-Doping Agency (WADA) and athletes have been disqualified and banned from the sport for using these drugs.

KEY POINTS

Muscle structure and function

1. Skeletal muscle cells are long, striated, multinucleated fibres. Skeletal muscles are attached to the bones, are striated, and can be controlled voluntarily, allowing movement and maintenance of posture.

2. Myofibrils are the contractile elements, composed of chains of sarcomeres containing thin (actin) and thick (myosin) filaments arranged in a regular array. The heads of the myosin molecules form cross-bridges that bind reversibly to the actin filaments, causing the filaments to slide over each other toward the centres of the sarcomeres.

3. Regulation of skeletal muscle contraction involves release of calcium from the sarcoplasmic reticulum following transmission of an action potential along the sarcolemma. Calcium initiates cross-bridge activity and sliding of the filaments. Cross-bridge activity ends when calcium ions are pumped back into the sarcoplasmic reticulum.

4. There are three types of muscle fibres, classified according to their contractile speed and metabolic characteristics as Type-I (slow-twitch oxidative), Type-IIa (fast-twitch oxidative), and Type-IIX (fast-twitch glycolytic). Most muscles contain a mixture of fibre types.

Proteins: structural and functional characteristics

1. Proteins, which consist of one or more amino acid sequences, are a major structural component of the body, but also have functional properties, acting as enzymes, hormones, receptors, and transporters, as well as making up the contractile apparatus of muscle.

2. All amino acids have a common structure, consisting of amino and carboxyl groups linked to a single carbon atom. Also present is an organic side chain, and the different side groups give the 20 amino acids their different identities. Some amino acids cannot be synthesized by animal tissues, although they are synthesized by plants, and these amino acids must be supplied from the diet.

3. Proteins consist of linear chains of amino acids linked by peptide bonds. Folding of these chains occurs because of interactions between the side groups of the component amino acids. The ultimate shape of the protein molecule and its ability to change shape give it its functional characteristics.

4. Synthesis of proteins is controlled by the genetic information contained in DNA. This determines the sequence of amino acids in a protein chain, as well as initiating and terminating the process. Changes in

KEY POINTS *(CONTINUED)*

the body's structure and function are brought about by changing the extent to which the genetic potential is expressed.

5. Although the body's structure is stable, there is a continuous process of protein synthesis and degradation: the half-life of individual proteins varies from less than one hour to several weeks. This determines the rate of adaptation to environmental stimuli, including exercise training. Muscle contains 40% of the total protein in a human body and accounts for 30–50% of all protein turnover in the body. The contractile proteins actin and myosin are the most abundant proteins in muscle, together accounting for 65% of all muscle protein.

6. The carbon skeleton of some amino acids can be used as a fuel for oxidative metabolism or can be used for the synthesis of other compounds. Proteins are not a major fuel for energy production during exercise, but regular resistance training will increase the dietary protein requirement.

7. After resistance exercise, muscle protein turnover is increased because of an acceleration of both protein synthesis and degradation. Muscle protein breakdown is increased after resistance exercise but to a smaller degree than muscle protein synthesis. The elevations in protein degradation and synthesis are transient but are still present at 24 h after exercise, although protein turnover returns to baseline levels after 48 h.

Proteins as enzymes

1. Enzymes are globular proteins that increase the rate of specific biochemical reactions: they act as catalysts. Their actions can be regulated in a way that permits the body to control the interactions between the different metabolic pathways (a sequence of several enzyme-catalysed reactions) that sustain life.

2. The activity of enzymes can be assessed by measuring the rate of substrate utilization or product formation under standardized conditions. The most common unit of measurement is the international unit (U or IU). This is the amount of enzyme that will convert 1 μmole of substrate to product in 1 min under the conditions (pH, temperature) that are specified for that reaction. The SI unit is the katal (kat), which is defined as the amount of enzyme that will convert 1 mole of substrate to product in 1 s under optimum conditions.

3. Many enzymes require the presence of one or more coenzymes if the reaction is to proceed: these are compounds that participate in the reaction, and are chemically altered. The conversion of pyruvate to lactate, for example, requires that NADH and H^+ are available and the NADH is oxidized to NAD^+. Some coenzymes such as NAD^+ are loosely bound to the enzyme, but others (e.g. biotin, haem) are tightly bound and are referred to as prosthetic groups.

4. Enzymes function by acting as reusable catalysts: this involves the formation of an enzyme–substrate complex as an intermediate step in the reaction, which lowers the energy of activation. Because less energy now has to be added, the reaction is more likely to proceed. Although the enzyme participates in the reaction it is not consumed and is therefore required to be present in only small amounts.

5. Enzyme activity increases with temperature, but at high temperatures (above about 40–45 °C) will fall sharply and irreversibly due to structural changes caused by denaturation of the protein. The local pH may affect the ionization of the substrate and of the enzyme's amino acid side chains and so enzyme activity is sensitive to changes in pH.

6. Allosteric modulation of enzyme activity refers to the reversible binding of small molecules to the enzyme at sites other than the active (substrate-binding) site. This produces a conformational shape change in the enzyme molecule that may increase or decrease its activity. Covalent modulation involves phosphorylation or dephosphorylation of an enzyme—usually involving the hydroxyl (—OH) group of a serine residue in the polypeptide chain. Allosteric and covalent modulation are important means of regulating enzyme activity in the cell.

7. The initial reaction rate $(V_0) = V_{max}[S]/(K_m + [S])$, where V_{max} is the maximum reaction rate (a function of the enzyme concentration present), $[S]$ is the substrate concentration, and K_m is the Michaelis constant, which is the substrate concentration at which the reaction rate is half the maximum value. K_m is a measure of the affinity of the enzyme for its substrate. Low K_m = high affinity.

Energy for muscle contraction

1. During any biochemical reaction free energy is released that can then be used to do work. In muscle, free energy released during the combustion of carbohydrates and lipids can be stored as chemical energy in the form of ATP. Free energy released by the subsequent degradation of ATP in the adenylate kinase reactions can be converted into mechanical energy to produce work (muscle can perform 24 kJ of work for each mole of ATP degraded). ATP is the only compound in living cells that enables this energy conversion to occur.

2. The energy source for all forms of muscle contraction is ATP, which is continuously regenerated during exercise from phosphocreatine, anaerobic metabolism of glycogen or glucose, or aerobic metabolism of acetyl-CoA derived principally from breakdown of carbohydrate or fat.

3. For rapid, high-force contractions, as used in weightlifting, the main source of energy is from the small ATP store in muscle.

Nutritional effects on strength training and performance

1. There is considerable controversy about the optimal protein intake for athletes. Although it is generally accepted that protein requirements are increased, the extent to which they are increased is subject to continuing debate.

2. It has been estimated that protein can contribute up to about 15% to energy expenditure in resting conditions. During exercise this relative contribution is likely to decrease because of an increasing importance of carbohydrate and fat as fuels. During very prolonged exercise when carbohydrate availability becomes limited, the contribution of protein to energy expenditure may amount to about 10% of total energy expenditure.

3. After each resistance training session, the rates of both protein synthesis and protein breakdown in the exercise muscles are increased. Training in the fasted state will limit the gains in muscle mass that can be made. The timing and composition of food intake in relation to exercise may be important to maximize adaptation to training.

4. The recommended protein intake for strength athletes is generally 1.6–1.7 g/kgbw/day, about twice the value for the average population. With increasing food intake the intake of protein automatically increases because many food products contain at least some protein. There is usually a linear relationship between energy intake and protein intake.

5. Excessive protein intake (more than 3 g/kgbw/day) may have various negative effects including kidney damage, increased blood lipoprotein levels (which have been associated with the development of arteriosclerosis), and dehydration.

6. Arginine and ornithine, in high doses, can stimulate release of growth hormone, but exercise itself has a greater effect. Thus, supplementation of the diet with individual amino acids is very unlikely to be effective in building muscle.

7. Supplementing the diet with creatine can increase the creatine content of muscle, improve strength and power, and typically results in an increase in body mass of 1–2 kg.

Selected further reading

Astrand P-O and Rodahl K (1986). *Textbook of work physiology*, 3rd edn. New York: McGraw-Hill.

Andersen LL *et al.* (2005). The effect of resistance training combined with timed ingestion of protein on muscle fiber size and muscle strength. *Metabolism* 54, 151–156.

Brooks GA and Fahey TD (1984). *Exercise physiology, human bioenergetics and its applications*. New York: John Wiley.

Burd NA *et al.* (2009). Exercise training and protein metabolism: influences of contraction, protein intake, and sex-based differences. *Journal of Applied Physiology* 106: 1692–1701.

Clarkson PM and Rawson ES (1999). Nutritional supplements to increase muscle mass. *Critical Reviews in Food Science and Nutrition* 39: 317–328.

Cornish-Bowden A and Wharton CW (1988). *Enzyme kinetics*. Oxford: IRL Press.

Greenhaff PL and Hultman E (1999). The biochemical basis of exercise. In: *Basic and applied sciences for sports medicine* (ed. Maughan RJ). Oxford: Butterworth-Heinemann.

Hames BD, Hooper NM, and Houghton JD (1997). *Instant notes in biochemistry*. Oxford: BIOS Scientific Publishers.

Jeukendrup AE and Gleeson M (2010) *Sport nutrition*. 2nd edn. Champaign: Human Kinetics.

King DS *et al.* (1999). Effect of oral androstenedione on serum testosterone and adaptations to resistance training in young men. *Journal of the American Medical Association* 281: 2020–2028.

Koopman R *et al.* (2005). Combined ingestion of protein and free leucine with carbohydrate increases postexercise muscle protein synthesis in vivo in male subjects. *American Journal of Physiology (Endocrinology and Metabolism)* 288: 645–653.

McPartland A and Segal IH (1986). Equilibrium constants, free energy changes and coupled reactions: concepts and misconcepts. *Biochemical Education* 14: 137–141.

Marieb EN (1993). *Human anatomy and physiology*, 2nd edn. Redwood City: Benjamin Cummings.

Matthews CK, van Holde KE, and Ahern KG (ed.) (1999). *Biochemistry*. San Francisco: Benjamin Cummings.

Maughan RJ, Gleeson M, and Greenhaff PL (1997). *Biochemistry of exercise and training*. Oxford: University Press.

Moore DR *et al.* (2009). Differential stimulation of myofibrillar and sarcoplasmic protein synthesis with protein ingestion at rest and after resistance exercise. *Journal of Physiology* 587: 897–904.

Nissen SL and Sharp RL (2003). Effect of dietary supplements on lean mass and strength gains with resistance exercise: a meta-analysis. *Journal of Applied Physiology* 94: 651–659.

Phillips SM (2006). Dietary protein for athletes: from requirements to metabolic advantage. *Applied Physiology, Nutrition, and Metabolism* 31: 647–654.

Ottaway JH (1988). *Regulation of enzyme activity*. Oxford: IRL Press.

Rasmussen BB *et al.* (2000). Androstenedione does not stimulate muscle protein anabolism in young healthy men. *Journal of Clinical Endocrinology and Metabolism* 85: 55–59.

Rennie MJ (2005). Body maintenance and repair: how food and exercise keep the musculoskeletal system in good shape. *Experimental Physiology* 90: 427–436.

Snow R (2002). Skeletal muscle. In: *Physiological bases of sports performance* (ed. Hargreaves M and Hawley J). Sydney: McGraw-Hill Australia, pp. 9–26.

Stryer L (1988). *Biochemistry*, 2nd edn. San Francisco: WH Freeman.

Tarnopolsky M (2006). Protein and amino acid needs for training and 'bulking up'. In: Burke L, V Deakin (ed.) *Clinical sports nutrition* 3rd edn. McGraw Hill, Sydney, Chapter 4.

Tullson PC and Terjung RL (1991). Adenine nucleotide metabolism in contracting skeletal muscle. *Exercise and Sport Science Reviews* 19: 507–537.

Williams MH (1998). *The ergogenics edge*. Champaign, IL: Human Kinetics.

The sprinter

3

Learning objectives

After studying this chapter, you should be able to . . .

1. give a general description of the characteristics and pathways of anaerobic metabolism;

2. describe the roles of phosphocreatine and its contribution to energy supply in high-intensity dynamic exercise;

3. appreciate the relative contributions of phosphocreatine and glycogen breakdown to anaerobic ATP resynthesis during sprinting;

4. describe the concept of the cellular energy charge and explain why there is a loss of adenine nucleotides during very high intensity exercise;

5. discuss the causes of fatigue in sprinting;

6. describe the time course of phosphocreatine resynthesis following very high intensity exercise;

7. appreciate the nutritional needs to sustain sprint training.

Introduction

A sprinter relies predominantly but not exclusively on anaerobic metabolism

The sprinter has to sustain a very high power output over a relatively short period of time (usually between 3 and 20 s). As the intramuscular supply of ATP is sufficient to last only about 2 s there is a pressing need to resynthesize ATP extremely quickly and this is achieved by the breakdown of intramuscular stores of phosphocreatine and the rapid activation of glycolysis. Both of these processes occur without the utilization of oxygen; that is they are anaerobic means of regenerating ATP. However, sprinting is not entirely anaerobic! There is a contribution of carbohydrate oxidation to ATP resynthesis during sprinting that increases as the duration and distance of the sprint increases. The approximate contributions of aerobic and anaerobic metabolism in sprint running events are shown in Table 3.1. Note that the average speed of running during these events is far higher than the speed that would elicit 100% maximal oxygen uptake (VO_{2max}).

Anaerobic metabolism

Anaerobic metabolism allows ATP resynthesis without the use of oxygen

Human skeletal muscle can perform work in the absence of an adequate supply of oxygen as a consequence of its ability to generate energy anaerobically.

Distance (m)	Exercise duration (s)* and [Running speed (km/h)]†		% Aerobic	% Anaerobic
	Men	Women		
100	9.58 [37.4]	10.49 [34.3]	10	90
200	19.19 [37.5]	21.34 [33.6]	20	80
400	43.18 [33.3]	47.60 [30.3]	30	70

* Durations given are the current outdoor world records at 1 March 2010.

† Average running speed based on these current world record times. The peak running speed would be a little higher as the average running speed includes the initial acceleration phase and the slowing down due to fatigue during the final stages of the race.

Table 3.1 Approximate contribution of aerobic and anaerobic energy sources to total energy production in running events of different durations involving maximal work.

Oxygen is never completely absent, so tissues themselves are never in a truly anaerobic environment, but the term is applied to the generation of energy by reactions that do not involve oxygen. Two separate systems are available to the muscle to permit this, and these are the phosphagen system and the glycolytic system. Because the glycolytic system depends on the production of lactic acid, whereas the phosphagen system involves no lactate formation, these systems of anaerobic ATP regeneration are sometimes referred to as the lactic and the alactic systems, respectively. Note that the terms lactic acid and lactate are often used interchangeably, but although lactic acid is perhaps a more descriptive term, clearly indicating the acidic nature of the molecule, lactate is more accurate and is used here.

Phosphagen system

ATP and PCr are referred to as the phosphagens

Much of our knowledge about muscle energy metabolism has come from the study of isolated animal muscles made to contract by electrical stimulation of the motor nerve or of the muscle directly. In such isolated muscle preparations, if a muscle is poisoned with cyanide (preventing oxidative phosphorylation in the mitochondria) and iodoacetic acid (inhibiting glycolysis), so that it can derive no energy from oxidative metabolism, or from the production of lactate, it can still contract strongly for a short period of time before fatigue occurs. This tells us that the muscle has another source of energy, and also that the capacity of this energy source is limited. This source is the intramuscular store of ATP (see Chapter 2) and phosphocreatine (PCr, also known as creatine phosphate): together, ATP and PCr are referred to as the phosphagens. The most important property of the phosphagens is that the energy store they represent is available to the muscle almost immediately.

ATP use in very high-intensity exercise

The muscle ATP concentration declines during very high-intensity exercise

Unlike the situation in prolonged submaximal exercise, the intramuscular ATP concentration does decline significantly during 10–60 s of all-out sprinting. During 10 s of maximal exercise the ATP concentration in Type-I fibres is found to be unchanged, but in Type-IIa fibres the ATP concentration falls by about 40%, and in Type-IIX fibres by about 70% (see Figure 3.1). If maximal effort exercise is maintained for a further 15 s, then the ATP concentration in Type-II fibres falls a little further and there is a significant drop (but only by about 20%) in the

Figure 3.1

(a) Decline in the ATP concentration for Type-I, IIa, and IIX fibres of the human vastus lateralis muscle during 10 and 25 s of maximal isokinetic cycling. (b) Power output (PO) profile during 25 s of maximal isokinetic cycling. Note that peak power output is reached at about 3 s; thereafter power declines (i.e. fatigue ensues).

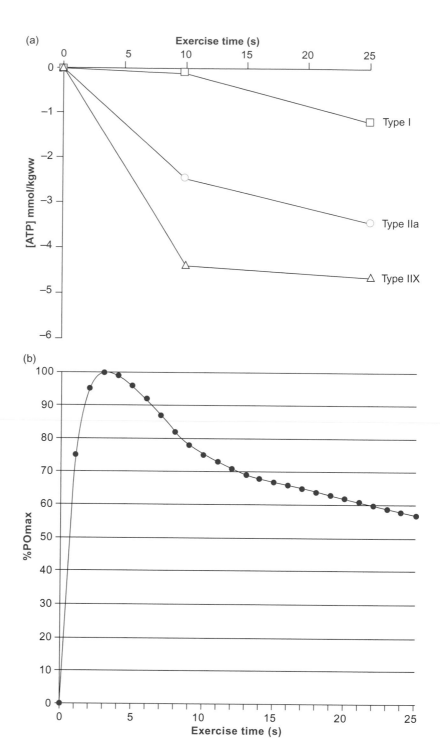

ATP concentration in Type-I fibres. In maximal exercise peak power is attained within 2–3 s and after 10 s there is typically a 20–25% loss of power output. These findings suggest that Type-II fibres contribute little to mechanical power output after the first 10 s of maximal exercise. Furthermore, it seems probable that the progressive muscle fatigue seen in a bout of all-out exercise is the consequence of a sequential failure of fibre-type populations in relation to their contractile and mechanical properties (i.e. the first fibres to fatigue are Type-IIX, then Type-IIa, followed by Type-I).

Phosphocreatine

Phosphocreatine can be used to resynthesize ATP at a very high rate

The PCr in muscle can be used to resynthesize ATP at a very high rate (considerably higher than glycolysis or the oxidative metabolism of carbohydrate or fat). This high rate of energy transfer corresponds to the ability to produce a high power output (power being the rate at which work is performed). The major disadvantage of this system is its limited capacity—the total amount of energy available is small (Table 3.2). If no other energy source is available to the muscle, fatigue will occur rapidly. During short bursts of running over a distance of 30–50 m, no slowing down occurs over the last few metres—full power can be maintained all the way—and the energy requirements are largely met by breakdown of the phosphagen stores. Over longer distances, running speed begins to fall off, as these stores decline and power output starts to fall. However, the rate of recovery from a short sprint is quite rapid, and a second burst can be completed at the same speed after only 2–3 min recovery. For longer sprints (100 m or more) much longer recovery periods are needed before the ability to produce a maximum performance is restored. These are important considerations for sports that involve multiple sprints in the course of a game, such as football, rugby, hockey, and basketball.

Figure 3.2 shows that at the onset of high-intensity exercise, the rates of PCr hydrolysis and lactate production are increased rapidly compared with rest. The greater the exercise intensity, the greater the rate of decline of PCr and accumulation of lactate. It is unclear whether these responses occur because of a lag in oxygen delivery and/or inertia in the activation of mitochondrial ATP resynthesis (TCA cycle and oxidative phosphorylation) at the onset of contraction. During brief all-out sprinting, the rate of ATP demand far exceeds the capacity of mitochondrial ATP resynthesis, and therefore anaerobic metabolism becomes the dominant contributor to ATP resynthesis. In physiological terms, this contribution from anaerobic metabolism to ATP resynthesis, whether at the onset of moderate intensity exercise or during high-intensity exercise, appears as the oxygen deficit.[1]

Table 3.2 Capacity and power of anaerobic systems for the production of ATP.

	Capacity (mmol ATP/kgdm)	Power (mmol ATP/kgdm/s)
Phosphagen system	80	9.0
Glycolytic system	300	4.5
Combined	380	11.0*

Values are expressed per kg dry mass (dm) of muscle [to convert to per kg wet weight (ww) simply divide by 4, as about 75% of muscle weight is water] and are based on estimates of ATP provision during high-intensity exercise of human vastus lateralis muscle.
* The combined maximum rate of ATP resynthesis is slightly lower than the sum of the maximum rates for the phosphagens (i.e. PCr breakdown) and glycolysis as these rates do not temporally coincide: the maximum rate of ATP resynthesis from PCr hydrolysis occurs in the first 1–2 s of exercise, whereas the maximum rate of ATP resynthesis from glycolysis is not reached until 5–10 s of exercise. Note that the maximum rates of ATP resynthesis from carbohydrate (glycogen) and fat oxidation are only about 2.8 and 1.0 mmol ATP/kgdm/s, respectively.

Figure 3.2

Changes in the intramuscular concentration of phosphocreatine, lactate, and ATP at different intensities of exercise.

Skeletal muscle contains about three to four times more phosphocreatine than ATP

Phosphocreatine is restricted to the cytoplasm of the muscle cell, where it is present at a concentration of about 20 mmol/kgww [80 mmol/kg dry matter (dm)]. Note that this is about 3–4 times higher than the intramuscular ATP concentration.

The structure of PCr is shown in Figure 3.3. Free creatine is present in resting skeletal muscle at a concentration of about 12–25 mmol/kgww (50–100 mmol/kgdm), but it is not synthesized in muscle tissue. Creatine is obtained from the diet. Because over 95% of the body's creatine is contained in skeletal muscle, meat is a very good source of creatine, providing about 1 g/day from a typical Western diet. The intramuscular creatine and PCr stores can be increased by dietary creatine supplementation; further consideration of this is given in Chapter 6.

Figure 3.3

Structure of phosphocreatine and creatine.

Leaders in the field

Otto Meyerhof was born on April 12, 1884, in Hannover. In 1909 he graduated in medicine, and developed a particular interest in psychology and philosophy. Otto Warburg, however, persuaded him to become interested in cell physiology, and he qualified in 1913 as a university lecturer in physiology. In 1929 he took charge of the newly founded Kaiser Wilhelm Institute for Medical Research at Heidelberg. In 1938 conditions became too difficult for him and he left Germany for Paris, but when the Nazis invaded France, he moved again and ended up in the United States where he began to investigate the energy changes in muscle. He proved that, in isolated but otherwise intact frog muscle, the lactic acid formed is reconverted to carbohydrate in the presence of oxygen, and that an extract of muscle could carry out all the steps of glycolysis with added glycogen and hexose-diphosphate in the presence of hexokinase derived from yeast. In this system, glucose was also metabolized and this was the foundation of the Embden–Meyerhof theory of glycolysis. For his discovery of the fixed relationship between the consumption of oxygen and the metabolism of lactic acid in the muscle, Meyerhof was awarded the Nobel Prize for Physiology or Medicine for 1922 jointly with AV Hill. The discovery of Meyerhof and his students that some phosphorylated compounds are rich in energy led to a revolution in the understanding of muscular contraction and of the entire significance of cellular metabolism. In 1944 he suffered a heart attack, and in 1951 another one that ended his life.

Creatine transport into muscle is against the concentration gradient and is coupled to that of sodium

Creatine is also synthesized in the liver from several amino acids (Figure 3.4). Following its release into the circulation, creatine is taken up into muscle. Because the concentration of creatine is far greater in muscle than in the blood plasma, there is a tendency for creatine to leak out of the muscle by simple diffusion. Energy is required to transport the creatine across the sarcolemmal membrane from the plasma into the muscle against this concentration gradient. This type of membrane transport is called active transport and the requirement for energy is common to the transmembrane transport of all substances against their prevailing concentration gradients. For charged particles, the electrical gradient also has to be taken into account, because excitable cells such as muscle possess a resting membrane potential that makes them electrically negative on the inside compared to the outside. This electrical potential difference is usually about 70 mV. In the case of creatine its transport is coupled to that of sodium; the large concentration difference for sodium across the membrane (approximately 140 mmol/l in the extracellular fluid compared with only 12 mmol/l inside the cell) is set up by the activity of the 'sodium pump', otherwise known as the sodium–potassium ATPase, which exports three sodium ions out of the cell for every two potassium ions that enter the cell at the expense of ATP breakdown to ADP and P_i.

The free energy of PCr hydrolysis is greater than that of ATP

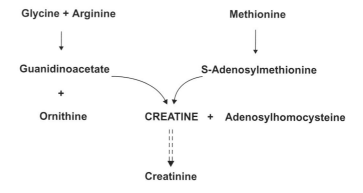

Figure 3.4

Biosynthesis of creatine from the amino acids arginine, glycine, and methionine. The first step in creatine synthesis involves the reversible transfer of an amidine group from arginine to glycine to form guanidinoacetic acid. This is followed by an irreversible transfer of a methyl group from S-adenosylmethionine to guanidinoacetic acid, forming creatine. This pathway occurs in the liver and kidneys. In muscle, creatine is broken down to creatinine, which is excreted in the urine.

The rapid degradation of PCr at the onset of moderate-intensity exercise and during high-intensity exercise occurs because it has a higher phosphate group transfer potential than ATP. This means that the free energy of PCr hydrolysis (−43 kJ/mol) is greater than that of ATP hydrolysis (−31 kJ/mol), resulting in a greater likelihood for free energy transfer to occur from PCr to ADP to re-form ATP:

$$ADP + PCr + H^+ \leftrightarrow ATP + Cr.$$

It can be seen, therefore, that this reaction functions to maintain ATP homeostasis during contraction at the expense of PCr. Note that the resynthesis of ATP via breakdown of PCr buffers some of the hydrogen ions formed as a result of ATP hydrolysis. The PCr in muscle is immediately available at the onset of exercise and can be used to resynthesize ATP at a very high rate.

Indeed, the rate at which this reaction can occur is far in excess of any of the ATP-utilizing reactions occurring in the cell, and it is not unusual for the muscle PCr store to be almost completely degraded during maximal exercise.[2] This reaction is termed the creatine kinase reaction because it is catalysed by the enzyme creatine kinase. Note that the reaction is reversible: depending on the energy state of the cell, it can go in either direction. During recovery from exercise, when ATP is regenerated from oxidative phosphorylation, creatine kinase can use ATP to replenish the PCr store.

Phosphocreatine appears to have multiple roles in muscle

It is now clear that creatine kinase has a number of isoenzymes (variations of the enzyme, having a slightly different structure but the same substrate specificity) that are located at different intracellular locations. At least three are known to be present in skeletal muscle. For example, MM-CK is located near the sites of muscle cross-bridge formation (i.e. near a site of ATP utilization) and Mi-CK is located at the mitochondrial membrane (i.e. near the site of ATP production). The discovery of the existence of isoenzymes of creatine kinase with discrete cellular locations has led to the hypothesis that PCr may have a number of different functions within skeletal muscle. The first, and possibly the most important, relates to its function described above, that is acting as a temporal buffer to maintain the cellular ATP concentration and the ATP to ADP ratio.

A second function, which is currently the subject of much debate, is that PCr may act as a spatial energy buffer, that is an energy transport system between the site of ATP production (the mitochondria) and the sites of ATP utilization (e.g. the myofibrils). This suggested function is illustrated in Figure 3.5 and has resulted in the use of the phrase 'the PCr shuttle'. Those researchers in favour of its existence have gone on to suggest that the primary role for PCr in Type-I muscle fibres may be to operate as a spatial buffer, which contrasts with its suggested principal role in Type-II fibres as a temporal energy buffer. The 3–5 mmol/kgww (12–20 mmol/kgdm) higher concentration of PCr found in Type-II fibres supports this suggestion.

Figure 3.5

The phosphocreatine shuttle. PCr, phosphocreatine; Cr, creatine; Mi-CK, mitochondrial creatine kinase; MM-CK, muscle (M isoform located at the M line of the sarcomere) creatine kinase. At the mitochondrial site, newly synthesized ATP enters the space between the inner and outer mitochondrial membranes where a portion is utilized by Mi-CK for the formation of PCr. The resulting ADP is transported into the mitochondrion. The PCr diffuses to the myofibrils where the CK located at the M line regenerates ATP from ADP formed during cross-bridge formation.

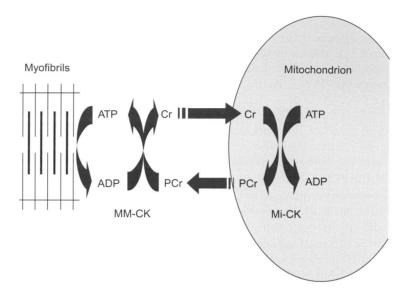

A third suggested function for PCr is its functional coupling with several other cellular reactions, which facilitates the integration of energy metabolism during muscle contraction. For example, it is clear from the reactions described above that the adenylate kinase reaction will result in the generation of H^+ ions and the creatine kinase reaction will result in the sequestering of H^+ ions; it is the functional coupling of these two reactions that prevents the rapid acidification of the cell at the onset of contraction. Similarly, the rapid liberation of P_i by ATP hydrolysis during contraction plays an integral part in the activation of glycogen phosphorylase at the onset of exercise, thereby ensuring that energy production is maintained. Increased intracellular concentrations of ADP and AMP are also important in this regard.

 Techniques

Magnetic Resonance Imaging (MRI) and Magnetic Resonance Spectroscopy (MRS). MRI is a medical imaging technique that is now widely used in medicine to obtain images of the internal structure and function of the body. Unlike other imaging techniques such as X-rays or computed tomography (CT) MRI uses no ionizing radiation: instead, it uses a strong magnetic field to align atoms of hydrogen that are present in water in the body. Radio-frequency (RF) fields are used to systematically alter the alignment of these atoms, causing the hydrogen nuclei to produce a rotating magnetic field that is detected by the scanner. MRS allows the collection of detailed biochemical information about metabolic processes by picking up signals from different chemical nuclei within the body. Apart from hydrogen, the most common nuclei to be studied are phosphorus, sodium, and fluorine. With these markers, the localized concentrations of a whole range of compounds of interest can be measured non-invasively.

Metabolic response to very high-intensity exercise

ATP resynthesis from phosphocreatine breakdown

> Phosphocreatine breakdown is initiated immediately at the onset of contraction

During high-intensity exercise, the relatively low rate of ATP resynthesis from oxidative phosphorylation results in the rapid activation of anaerobic energy production from both PCr and glycogen hydrolysis. PCr breakdown is initiated immediately at the onset of contraction to prevent the rapid accumulation of ADP resulting from ATP hydrolysis. However, as you can see in Figure 3.6, the rate of PCr hydrolysis begins to decline after only a few seconds of very high force generation. The importance of PCr hydrolysis lies in the extremely rapid rates at which it can resynthesize ATP. This is especially true of maximal short-duration exercise. For example, Figure 3.6 shows the rate of muscle ATP resynthesis from PCr hydrolysis during 30 s of maximal fatiguing isometric contraction. First, note that PCr utilization is at its highest within 2 s of the initiation of contraction. Secondly, however, you can see that after only 2.6 s of contraction the ATP yield from PCr is reduced by about 15%, and following 10 s of contraction it is reduced by more than 50%. The contribution of PCr to ATP resynthesis in the last 10 s of a 30-s exercise bout is relatively small, amounting to only 2% of the initial yield. By this time, of course, the actual force (or power) will also have declined substantially. As the rate of PCr breakdown declines, so does the rate of ATP resynthesis because the other ATP-regenerating mechanisms (glycolysis and oxidative phosphorylation) cannot resynthesize ATP at as fast a rate. The decreased turnover of ATP in the muscle means that the force or power must also fall.

Because PCr is so important to muscle performance during short bursts of activity, it is tempting to ask why the body does not store more PCr in its muscles. In the wild, many animals need to be able to move quickly to catch prey or to escape from being eaten themselves. The requirement to develop athletic prowess in the animal kingdom has been a strong driver of evolution, but no animals seem to have adapted by greatly increasing the creatine or PCr contents of their muscles. The most likely reason for this is the weight penalty that would be incurred. PCr is a relatively small molecule and to increase its concentration in the muscle would have an osmotic effect, retaining more water in the muscle and therefore increasing body mass. Any improvement in muscle performance would be offset by the increased inertia and energy requirement to move a heavier body.

The mechanisms responsible for the almost instantaneous decline in the rate of PCr utilization during maximal exercise are at present unknown, but may

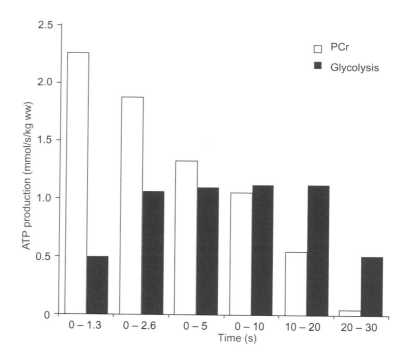

Figure 3.6

Rates of anaerobic ATP resynthesis from phosphocreatine (PCr) and glycolysis during 30 s of near maximal intensity isometric contraction in man. Values were calculated from metabolite changes measured in biopsy samples obtained during intermittent electrically evoked contraction (1.6 s stimulation at 50 Hz, 1.6 s rest).

be related to a local decline in the availability of PCr close to the cross-bridges where it is needed. Considering the high energy demand of maximal exercise, it is possible that the very rapid rate of PCr utilization at the onset of contraction could be responsible for a rapid depletion of stores at the sites of rapid energy translocation (actomyosin cross-bridges). Certainly, this seems plausible, because when intense exercise is continued for more than 20 s, the cellular store of PCr is almost completely depleted, probably as a consequence of mitochondrial ATP production being unable to match the rate of PCr hydrolysis. However, it should be borne in mind that even when the PCr store is almost zero, muscle can continue to function, albeit at a much reduced power.

ATP resynthesis from glycogen metabolism

Glycogen breakdown and glycolysis are rapidly activated within the first few seconds of intense exercise

If high-intensity exercise is to continue beyond only a few seconds there must be a marked increase in the contribution from glycolysis to ATP resynthesis. Glycogenolysis is the hydrolysis of muscle glycogen to glucose 1-phosphate and glycolysis is the series of reactions involved in the degradation of glucose 1-phosphate to lactate (details of this pathway are given in Chapter 4). The integrative nature of energy metabolism ensures that the activation of muscle contraction by Ca^{2+} and the accumulation of the products of ATP and PCr hydrolysis (ADP, AMP, IMP, NH_3, and P_i) act as stimulators of glycogenolysis and glycolysis, and in this way guarantee that anaerobic ATP production is maintained, at least in the short term.

Anaerobic glycolysis involves several more steps than PCr hydrolysis, and can provide ATP at a slower rate, but compared with oxidative phosphorylation is still very rapid. As described in some detail in Chapter 4, the generation of ATP in glycolysis occurs via the phosphorylation of ADP in the second half of the pathway. It was thought for many years that PCr was the sole fuel used at the initiation of contraction, with glycogen utilization occurring only when the PCr concentration had become depleted. This is now known not to be the case. As you can see in Figure 3.6, ATP resynthesis from glycolysis during 30 s of maximal fatiguing contraction begins almost immediately at the onset of exercise. Furthermore, unlike PCr hydrolysis, ATP production from glycolysis does not reach its maximal rate until after 5 s of exercise and is maintained at this high rate for several seconds. Over 30 s of exercise, the contribution from anaerobic glycolysis to ATP resynthesis is nearly double that from PCr.

> Glycogen catabolism will supply the major part of the energy requirement for maximum-intensity efforts lasting from 20 s to about 2 min

In sprinting, the muscle glycogen stores are broken down rapidly with a correspondingly high rate of lactate formation: some of the lactate diffuses out of the muscle fibres where it is produced and appears in the blood. A substantial proportion, amounting to about 25 mmol glucosyl units/kgww (100 mmol/kgdm), of the muscle glycogen store can be used for anaerobic energy production during high-intensity exercise, and will supply the major part of the energy requirement for maximum-intensity efforts lasting from 20 s to about 2 min (see Chapter 4). Because 3 mmol of ATP can be resynthesized by anaerobic glycolysis from each mmole of glucosyl units derived from the breakdown of glycogen, the capacity for ATP regeneration from anaerobic glycogen breakdown is about 75 mmol ATP/kgww (300 mmol/kgdm), which is three to four times greater than that available from complete hydrolysis of the muscle PCr store. For sprints lasting less than 20 s, the phosphagens are the major energy source. Although the total capacity of the glycolytic system is greater than that of the phosphagen system, the rate at which it can produce energy (i.e. resynthesize ATP) is lower (Table 3.2). The power output that can be sustained by this system is therefore correspondingly lower, and it is for this reason that maximum speeds cannot be sustained for more than a few seconds; once the phosphagens are depleted, the rate of work output must necessarily fall. As mentioned in Chapter 2, Type-II fibres have a higher content of PCr and glycogen than Type-I fibres. Type-II fibres also possess a greater amount of phosphorylase, the enzyme that breaks down glycogen. It is not surprising, therefore, that biopsy studies have shown a higher proportion of Type-II fibres in elite sprinters than in endurance athletes and the sedentary population.

Phosphocreatine and glycogen breakdown in Type-I and Type-II fibres

Rates of phosphocreatine and glycogen breakdown during very high intensity exercise are faster in Type-II muscle fibres

Most of the conclusions presented so far have been based on metabolite changes measured in studies of isolated animal muscle and from human studies in which biopsy samples were obtained from the quadriceps femoris muscle group at the front of the thigh. However, human skeletal muscle is composed of at least two functionally and metabolically different fibre types (as described in Chapter 2). Type-I fibres are characterized as being slow contracting, fatigue resistant, capable of a relatively low power output and favouring aerobic metabolism for ATP resynthesis during contraction. Conversely, Type-II fibres are relatively fast contracting, fatigue rapidly, are capable of a high power output, and favour mainly anaerobic metabolism for ATP resynthesis. Evidence from animal studies that have utilized muscles composed of predominantly Type-I or Type-II fibres suggests that the rapid and marked rise and subsequent decline in maximal power output observed during intense muscle contraction may be closely related to activation and rapid fatigue of Type-II fibres. Similar evidence for humans is not readily available because human limb muscles tend to have a more mixed fibre-type composition, although somewhat similar results have been reported from one study using bundles of similar human muscle fibre types.

Look at Figure 3.7, which shows the rates of PCr and glycogen degradation in Type-I and Type-II muscle fibres during maximal exercise under three different experimental conditions. Note, first, that at rest PCr and glycogen concentrations are higher in Type-II muscle fibres than in Type-I fibres, and second, that during intense contraction the rates of glycogenolysis and PCr degradation are higher in Type-II than in Type-I fibres. This is true for both dynamic exercise (treadmill sprinting) and electrically induced isometric contractions, which indicates that this response is not a function of the way in which the muscle is activated, but rather is a characteristic of the fibres themselves. The rates of glycogenolysis observed in both fibre types during treadmill sprinting and intermittent isometric contraction with circulation occluded are in good agreement with the maximal activity (V_{max}) of phosphorylase measured in both fibre types, suggesting that glycogenolysis is occurring at a near maximal rate during intense exercise.

Surprisingly, during repeated isometric contractions (each lasting 1.6 s) with circulation intact, when the rest interval between contractions is also 1.6 s, the rate of glycogenolysis in Type-I fibres is almost negligible. The corresponding rate in Type-II fibres is almost maximal and similar to that seen during contraction with circulatory occlusion. This suggests that during maximal exercise glycogenolysis in Type-II fibres is invariably occurring at close to the maximum rate, irrespective of the experimental conditions, while the rate in Type-I fibres

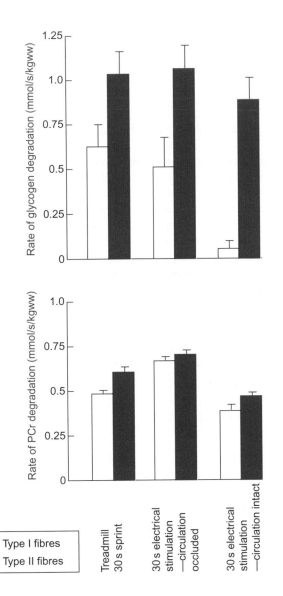

Figure 3.7

Rates of glycogen and phosphocreatine (PCr) degradation in Type-I and II muscle fibres during 30 s of maximal treadmill sprinting and 30 s of intermittent electrical stimulation (1.6 s stimulation at 50 Hz, 1.6 s rest) with circulation occluded and intact.

is probably very much related to cellular oxygen availability and phosphorylation potential or energy charge of the cell (see following sections).

Myokinase reaction

The myokinase reaction uses two molecules of ADP to generate one molecule of ATP

An additional pathway to regenerate ATP when ATP and PCr stores are depleted is through a kinase reaction[3] that utilizes two molecules of ADP to generate one molecule of ATP (and one molecule of adenosine monophosphate, AMP). This reaction is catalysed by the enzyme called myokinase, which is also known as adenylate kinase:

$$ADP + ADP \rightarrow ATP + AMP - 31 \text{ kJ/mol ADP.}$$

This reaction becomes important only during exercise of high intensity. Even then, the amount of energy it makes available in the form of ATP is extremely limited and the real importance of the reaction may be in the formation of AMP, which is a potent allosteric activator of a number of enzymes involved in energy metabolism, particularly those involved in glycogen breakdown and glycolysis.

Oxidative metabolism also makes a contribution to ATP resynthesis in a 10-s sprint. Although small (probably contributing less than 10% of total ATP resynthesis), it is still important, and, of course, the contribution from carbohydrate oxidation becomes increasingly greater as the duration of the bout of high-intensity exercise gets longer.

Loss of adenine nucleotides

Adenine nucleotide loss may be of importance to muscle function during conditions of metabolic crisis

It is known that the total adenylate pool can decline rapidly if the AMP concentration of the cell begins to rise during muscle force generation. This decline occurs principally via deamination of AMP to inosine monophosphate (IMP), but also by the dephosphorylation of AMP to adenosine. The loss of AMP may initially appear to be counterproductive because of the reduction in the total adenylate pool. However, it should be noted that the deamination of AMP to IMP only occurs under low ATP/ADP ratio conditions and, by preventing excessive accumulation of ADP and AMP, enables the adenylate kinase reactions to continue, resulting in an increase in the ATP/ADP ratio and continuing muscle force generation. Furthermore, it has been proposed that the free energy of ATP hydrolysis will decrease when ADP and P_i accumulate, which could further impair muscle force generation. For these reasons, adenine nucleotide loss has been suggested to be of importance to muscle function during conditions of metabolic crisis; for example during very high intensity exercise such as sprinting and in the later stages of prolonged submaximal exercise when glycogen stores become depleted.

The cellular energy charge and the adenylate pool

> The energy charge is a good indicator of the energy status of the cell

The concentrations of ATP, ADP, and AMP can be used to calculate the energy charge of the cell. This concept was proposed by Atkinson in 1977 (see recommended reading) and it is a measure of the extent to which the total adenine nucleotide pool of the cell (ATP, ADP, and AMP) is phosphorylated. It is described by the following equation:

$$\text{Energy charge} = \{[\text{ATP}] + 0.5\,[\text{ADP}]\}/\{[\text{ATP}] + [\text{ADP}] + [\text{AMP}]\}.$$

The energy charge is a good indicator of the energy status of the cell (i.e. its capacity to do work). For example, the energy charge of the cell would be 1.0 if the whole of the adenine nucleotide pool was in the form of ATP and under these conditions the cell would have a maximum energy charge. Conversely, the energy charge will be zero when ATP has been completely hydrolysed to AMP. Both these scenarios should only be viewed as theoretical examples, as the concentration of ATP in living human skeletal muscle will not decline by more than 60%, even during maximal exercise with blood flow completely occluded, and the adenylate nucleotide pool is never all in the form of ATP. Under normal resting conditions, the energy charge of skeletal muscle is in the region of 0.90 to 0.95. However, this has been shown to decline to around 0.85 in some disease states, and can fall to less than 0.7 during fatiguing high-intensity exercise. It should be noted that further falls are likely to be associated with irreversible cellular damage.

> ATP, ADP, and AMP act as allosteric activators or inhibitors of the key enzymatic reactions involved in energy metabolism

The rate of ATP resynthesis during exercise is regulated by the energy charge of the muscle cell. For example, the decrease in the energy charge at the onset of contraction, that is the momentary decline in ATP and increases in ADP and AMP, accelerates both anaerobic and oxidative ATP resynthesis, with the net effect of increasing the rate of energy supply to match the increased demand. If the energy charge continues to decline, ATP degradation will be inhibited, that is the muscle will fatigue and work output will fall. The relatively low concentration of ATP (and ADP) inside the cell means that any increase in the rate of hydrolysis of ATP (e.g. at the onset of exercise) produces a rapid change in the ratio of ATP to ADP and also increases the intracellular concentrations of AMP and P_i. These changes, in turn, activate enzymes that immediately stimulate the breakdown of intramuscular fuel stores to provide energy for ATP resynthesis. In this way, energy metabolism increases rapidly following the start of exercise.

ATP, ADP, and AMP act as allosteric activators or inhibitors of the key enzymatic reactions involved in PCr, carbohydrate (CHO), and fat degradation and utilization. For example, as already mentioned, creatine kinase, the enzyme responsible for the rapid rephosphorylation of ATP at the initiation of muscle force generation, is activated rapidly by an increase in cytoplasmic ADP concentration and is inhibited when the cellular ATP concentration is high. Similarly, glycogen phosphorylase, the enzyme that catalyses the conversion of glycogen to glucose 1-phosphate and thus primes the glycolytic pathway, is activated by increases in AMP and P_i (and calcium ion) concentration and is inhibited by an increase in ATP concentration.

Causes of fatigue in sprinting

Even in a 100-m sprint, the runners are slowing down slightly in the final third of the race

Fatigue has been defined as the inability to maintain a given or expected force or power output and is an inevitable feature of maximal exercise. Typically, the loss of power output or force production is likely to be in the region of 40–60% of the maximum observed during 30 s of all-out exercise. Measurements on elite sprinters performing all-out effort on a cycle ergometer show that maximum power is reached after only 3–4 s; thereafter power declines. During sprint running, the acceleration phase takes place over the first 4–5 s by which time the athlete will have covered about 30–40 m of the track. Thereafter, running speed declines, but not to the same extent as is seen on the cycle ergometer; this is due to the forward momentum that is generated when running fast, helping to keep the speed up. This explains why the time for the 200-m race is invariably less than double the time for the 100-m race. However, even in a 100-m sprint, the runners are slowing down slightly in the final third of the race. In other words, fatigue has already begun. Success in sprint events like the 100 and 200 m is therefore associated with the maximum running speed that can be achieved by the athlete and the ability to minimize the inevitable loss of power following the initial acceleration phase and attainment of maximum running speed.

The causes of fatigue in sprinting are multifactorial but the decline in phosphocreatine availability is probably the most important factor

Fatigue is not a simple process with a single cause; many factors can contribute to fatigue. However, during maximal effort exercise lasting less than 30 s, fatigue is caused primarily by a gradual decline in anaerobic ATP production or an increase in ADP accumulation caused by a depletion of PCr and a fall in the rate of glycolysis. In high-intensity exercise lasting 1–5 min, H^+ ion accumulation may contribute to the fatigue process (see Chapter 4). The general consensus at the moment seems to be that the maintenance of force production during very high-intensity exercise is pH dependent, but the initial force generation during the first few seconds of activity is more related to PCr availability.

> Accumulation of metabolites and altered calcium transport are also implicated in fatigue

One of the consequences of rapid ATP hydrolysis during high-intensity exercise is the accumulation of P_i, which has been shown to inhibit muscle excitation–contraction coupling directly. However, the depletion of PCr and the accumulation of P_i occur over a similar time course, which makes it difficult to separate the effect of PCr depletion from P_i accumulation *in vivo*. This problem is further confounded by the parallel increases in hydrogen and lactate ions that occur during high-intensity exercise. All of these metabolites have been independently implicated with muscle fatigue.

As described in Chapter 2, calcium release by the sarcoplasmic reticulum results from muscle depolarization and is essential for the activation of muscle contraction coupling. It has been demonstrated that during fatiguing contractions there is a slowing of calcium transport and progressively smaller calcium transients that has been attributed to a reduction in calcium reuptake by the sarcoplasmic reticulum. Strong evidence that a disruption of calcium handling is responsible for fatigue comes from studies showing that the stimulation of sarcoplasmic reticulum calcium release caused by the administration of caffeine to isolated muscle whose fibres have had their surface membrane removed can improve muscle force production, even in the presence of a low muscle pH. Alternatively, fatigue during high-intensity exercise may be associated with an excitation-coupling failure and possibly a reduced nervous drive due to reflex inhibition at the spinal level. In the latter hypothesis, accumulation of interstitial potassium in muscle may play a major role (for details see Sjogaard 1991; Bangsbo 1997).

Post-exercise recovery: the resynthesis of phosphocreatine

> PCr resynthesis after exercise follows an exponential curve and about half of it is restored in the first 30 s post-exercise

The creatine kinase reaction is an equilibrium reaction (as is the adenylate kinase reaction) and is therefore reversible:

$$ATP + Cr \leftrightarrow ADP + PCr + H^+.$$

Following exercise, when the energy charge of the cell is increased and sufficient free energy is available to rephosphorylate Cr, the reaction will proceed from left to right to restore muscle PCr levels.

In general, the resynthesis of PCr following complete degradation follows an exponential curve and the half-time for resynthesis (the time to resynthesize 50% of the resting store) is often quoted as 30 s (as illustrated in Figure 3.8). In reality, however, there appears to be a large variation in the time course of resynthesis depending on the type of exercise performed and the duration and number of exercise bouts completed. Factors known to influence the rate of PCr resynthesis during recovery from exercise are the cellular concentrations of ATP, ADP, and Cr, which is not surprising given the equilibrium nature of the creatine kinase reaction. In addition, the H^+ ion is known to be a potent inhibitor of creatine kinase. In practice, therefore, a low muscle pH, a low oxygen tension and/or a reduction in muscle blood flow will severely impair PCr resynthesis following exercise. Indeed, muscle ischaemia is often used as a tool in metabolic research to 'arrest' PCr resynthesis following muscle contraction, thereby providing sufficient time to enable relevant biochemical and physiological measurements to be made (see Figure 3.8).

The rate of phosphocreatine resynthesis is lower in Type-II fibres

It is now clear that there are differences in the rates of PCr resynthesis between muscle fibre types following exercise-induced PCr depletion. It seems that the rate of PCr resynthesis is significantly lower in Type-II muscle fibres during the first few minutes of recovery (possibly due to a greater fall in intracellular pH in this fibre type). After these initial few minutes, however, PCr resynthesis is accelerated in Type-II muscle fibres, such that after 15 min of recovery the concentration of PCr is in fact greater than that observed at rest. The mechanism responsible for this PCr overshoot in Type-II fibres or for how long this remains higher is currently unknown.

Following a bout of high-intensity exercise it takes considerably longer to remove the accumulated lactate and restore the muscle glycogen store than it does to fully resynthesize the PCr store. These other aspects of recovery are covered in subsequent chapters.

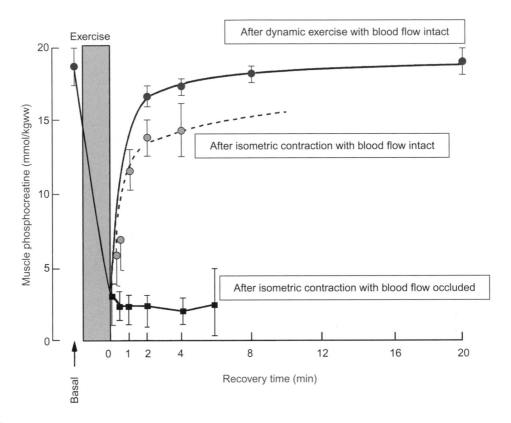

Figure 3.8

Time course of phosphocreatine resynthesis after maximal exercise.

Nutritional effects on sprint performance

Creatine supplements may help sprinters to sustain higher training intensities

Sprinters are concerned with generating high power outputs and building large and powerful muscles is a key goal. It is not surprising, therefore, that they share many training and nutrition practices with strength athletes. High-protein diets are common, but as with the strength athlete, these are probably often higher than is necessary for optimum performance. Creatine supplementation is also especially popular with sprinters: the first athletes to use this in major competition included some of the athletes who won gold medals in sprint events at the Barcelona Olympic Games in 1992. When creatine was first used, athletes would take it only in the last few days before competition to achieve a rapid boost of the phosphocreatine levels in their muscles. More often now, athletes use creatine supplementation over the whole season as a way of increasing the intensity of training that can be sustained. The reasoning is that training harder induces greater physiological and biochemical adaptations that in turn mean better performance in competition.

The obvious benefit of an increased creatine content in muscle (about two-thirds of the total is in the form of phosphocreatine) is an increased amount of immediately available high-energy phosphate groups that can be transferred to ADP. An increased creatine content, however, also means that there is a faster rate of phosphocreatine resynthesis after an intense sprint. This may be why the greatest performance-enhancing effects of creatine supplementation are generally seen when several short sprints are performed with insufficient time for complete recovery of phosphocreatine between sprints. This is typical of the interval training sessions carried out by many athletes in the longer sprint events and may explain why more intensive training can be achieved. It is also typical of the patterns of play in many team sports, and the use of creatine supplementation will be considered in this respect in more detail in Chapter 6.

The use of creatine has generated much controversy as it can be effective in improving performance. Although not all studies show a positive effect, the balance of the available evidence does support this, but its use is not prohibited in sport at the present time. Several long-term studies of creatine supplementation have now been undertaken, and there are no reports of adverse effects on health or performance from long-term use.

Performing repeated sprints places high demands on the muscle glycogen stores

Performance in sprints is less affected by the pre-exercise diet than is performance in prolonged exercise. Muscle glycogen availability *per se* is not usually considered to be responsible for fatigue during high-intensity exercise, providing the pre-exercise glycogen store is not depleted to below 25 mmol/kgww (100 mmol/kgdm). However, sprinters should be aware that performing repeated sprints in training will place high demands on the muscle glycogen stores. In a single 6-s all-out sprint on a laboratory treadmill, it was found that 16% of the glycogen present was broken down. Most of this was converted to lactate and some to the accumulated glycolytic intermediates, with only a small amount being oxidized (Gaitanos *et al.* 1993). After performing ten 6-s sprints (each separated by 30 s resting recovery) the muscle glycogen content had fallen by 40%. A typical training session for sprinters consists of several brief intense sprints, so it is clear that there will be a substantial decrease in the available muscle glycogen stores. Some of the lactate formed during the sprints is transported to the liver and used to resynthesize glucose, which can then be returned to the muscle for storage, but most is oxidized by the muscles during periods of low-intensity exercise between sprints. The body's substantial fat reserves cannot be converted to carbohydrate, so the diet must supply sufficient carbohydrate to allow replenishment of the glycogen stores between training sessions. The trend for sprinters and power athletes to eat a high-protein diet (which in practice often means that the intake of fat is also high) may mean that the dietary intake of carbohydrate is inadequate to allow an intensive training programme to be sustained.

KEY POINTS

1. Human skeletal muscle can exert force without the use of oxygen as a consequence of its ability to generate energy anaerobically. Two separate systems are available in the muscle to permit this: the phosphagens and the glycolytic pathway.

2. The phosphagens are the intracellular stores of ATP and phosphocreatine (PCr). The energy store they represent is available to the muscle almost immediately. The muscle only uses ATP as the direct source of energy for contraction, but the PCr in muscle can be used to resynthesize ATP at a very high rate. The major disadvantage of this system is its limited capacity—the total amount of energy available is small.

3. Phosphocreatine is present in the cytosol of muscle at about three times the concentration of ATP. It is generally accepted that the rapid degradation of PCr in the creatine kinase reaction at the onset of muscle force generation occurs because the free energy released can be used to resynthesize ATP, thereby maintaining a high cellular ATP to ADP ratio. However, the discovery of several isoenzymes of creatine kinase with defined cellular locations has led to the hypothesis that PCr may have a number of different functions within skeletal muscle.

4. The creatine kinase reaction is reversible and occurs following exercise when sufficient energy is available to rephosphorylate Cr. The resynthesis of PCr follows an exponential curve but the time course of resynthesis is dependent on a number of factors.

5. The sum of cellular ATP, ADP, and AMP concentrations is termed the total adenine nucleotide pool. The extent to which the total adenine nucleotide pool is phosphorylated is known as the energy charge of the cell, and it is a good indicator of the energy status of the cell. The rate at which ATP is resynthesized during exercise is known to be regulated by the energy charge of the muscle cell. For example, the decline in cellular concentration of ATP at the onset of muscle force generation and parallel increases in ADP and AMP concentrations (i.e. a decline in the energy charge) directly stimulate anaerobic and oxidative ATP resynthesis.

6. The total adenine nucleotide pool of the cell declines if the AMP concentration of the cell begins to rise during exercise. The loss of adenine nucleotides is potentially detrimental because it will reduce the availability of adenine nucleotides for phosphorylation. However, this adverse effect is outweighed in the short term by the stimulatory effect that the reduction in the cellular ADP and AMP concentrations has on the adenylate kinase reactions, resulting in an increase in the energy charge and continued force generation.

7. Skeletal muscle adenine nucleotide loss occurs first by the deamination of AMP to IMP and ammonia, and secondly by the dephosphorylation of AMP to adenosine. Both IMP and adenosine can be further degraded to inosine and then hypoxanthine, which then leaves muscle and is degraded and excreted by

KEY POINTS *(CONTINUED)*

the kidneys. The predominant pathway for adenine nucleotide loss in man is via deamination of AMP to IMP and ammonia; however, substantial variation is known to exist between muscle fibre types and animal species.

8. An alternative fate for IMP is that it can be used to resynthesize AMP. The deamination of AMP to IMP and subsequent reamination of IMP forms the purine nucleotide cycle. Several important roles have been proposed for the purine nucleotide cycle in muscle energy metabolism.

9. The close association between muscle adenine nucleotide loss and the development of fatigue during short-lasting intense exercise and prolonged submaximal exercise might suggest that AMP deamination is implicated in fatigue development. However, it is more likely that this association is a reflection of energy delivery failing to meet the energy demands of the exercise and that fatigue is due to a number of factors, including a local cellular depletion of phosphocreatine and ATP and an accumulation of ADP, P_i, and H^+ ions.

10. For exercise lasting more than a few seconds, ATP derived from the anaerobic metabolism of glucose (or glycogen) becomes available. Glycolysis is the name given to this pathway and the end-product of this series of reactions is pyruvate or lactate.

11. Glycolysis makes two molecules of ATP available for each molecule of glucose that passes through the pathway. If muscle glycogen is the starting substrate, three ATP molecules are generated for each glucose unit passing down the pathway.

12. Anaerobic glycolysis involves several more steps than PCr hydrolysis; however, compared with oxidative phosphorylation it is still very rapid. It is initiated at the onset of contraction, but unlike PCr hydrolysis does not reach a maximal rate until after 5 s of exercise and can be maintained at this level for several seconds during maximal muscle force generation. The mechanism(s) responsible for the eventual decline in glycolysis during maximal exercise have not been resolved.

13. The resynthesis of PCr following its complete degradation during a bout of very high-intensity exercise follows an exponential curve and the half-time for resynthesis (the time to resynthesize 50% of the resting store) is about 30 s.

14. Sprinters share many training and nutrition practices with strength athletes. High-protein diets are common, but as with the strength athlete, these are probably often higher than is necessary for optimum performance.

15. Creatine supplementation over the whole season may enable sprinters to increase the intensity of training that they can sustain.

16. Performance in sprints is less affected by the pre-exercise diet than is performance in prolonged exercise. Muscle glycogen availability *per se* is not usually considered to be responsible for fatigue during high-intensity exercise, providing the pre-exercise glycogen store is not depleted to below 25 mmol/kgww (100 mmol/kgdm). However, sprinters should be aware that performing repeated sprints in training will place high demands on the muscle glycogen stores.

Selected further reading

Atkinson DE (1977). *Cellular energy metabolism and its regulation*. New York: Academic Press.

Balsom PD *et al.* (1993). Creatine supplementation and dynamic high-intensity intermittent exercise. *Scandinavian Journal of Medicine and Science in Sports* 3: 143–149.

Bangsbo J (1997). Physiology of muscle fatigue during intense exercise. In: *The clinical pharmacology of sport and exercise* (ed. Reilly T and Orme M). Amsterdam: Elsevier.

Bangsbo J *et al.* (2000). Muscle oxygen kinetics at the onset of intense dynamic exercise in humans. *American Journal of Physiology* 279: R899–R906.

Casey A *et al.* (1996). Effect of creatine supplementation on muscle metabolism and exercise performance. *American Journal of Physiology* 271: E31–E37.

Gaitanos GC *et al.* (1993). Human muscle metabolism during intermittent maximal exercise. *Journal of Applied Physiology* 75: 712–719.

Green HJ (1986). Muscle power: fibre type recruitment, metabolism and fatigue. In: *Human muscle power* (ed. Jones NL, McCartney N, and McComas AJ). Champaign, IL: Human Kinetics, pp. 65–79.

Greenhaff PL *et al.* (1991). Energy metabolism in single human muscle fibres during contraction without and with epinephrine infusion. *American Journal of Physiology* 260: E713–E718.

Greenhaff PL *et al.* (1993). Energy metabolism in single human muscle fibres during intermittent contraction with occluded circulation. *Journal of Physiology* 460: 443–453.

Greenhaff PL *et al.* (1994). The metabolic responses of human type I and II muscle fibres during maximal treadmill sprinting. *Journal of Physiology* 478: 149–155.

Hultman E and Sjoholm H (1983). Substrate availability. In: *Biochemistry of exercise* (ed. Knuttgen HG, Vogel HG and Poortmans JA). Champaign, IL: Human Kinetics, pp. 63–75.

Maughan RJ, Gleeson M, and Greenhaff PL (1997). *Biochemistry of exercise and training*. Oxford: University Press.

McKenna MJ (2002). Mechanisms of muscle fatigue. In: *Physiological bases of sports performance* (ed. Hargreaves M and Hawley J). Sydney: McGraw-Hill Australia, pp. 79–107.

Medbo JI and Tabata I (1989). Relative importance of aerobic and anaerobic energy release during short-lasting exhausting bicycle exercise. *Journal of Applied Physiology* 67: 1881–1886.

Radford PF (1990). Sprinting. In: *Physiology of sports* (ed. Reilly T *et al.*). London: E & FN Spon, pp. 71–100.

Ren J-M and Hultman E (1989). Regulation of glycogenolysis in human skeletal muscle. *Journal of Applied Physiology* 67: 2243–2248.

Sahlin K and Broberg S (1990). Adenine nucleotide depletion in human muscle during exercise: causality and significance of AMP deamination. *International Journal of Sports Medicine* 11: S62–S67.

Sjogaard G (1991). Role of exercise-induced potassium fluxes underlying muscle fatigue: a brief review. *Canadian Journal of Physiology and Pharmacology* 69: 238–245.

Soderlund K *et al.* (1992). Energy metabolism in type I and type II human muscle fibres during short term electrical stimulation at different frequencies. *Acta Physiologica Scandivanica* 144: 15–22.

Spencer MR and Gastin PB (2001). Energy contribution during 200-to-1500 m running in highly trained athletes. *Medicine and Science in Sports And Exercise* 33: 157–162.

Spriet LL (1992). Anaerobic metabolism in human skeletal muscle during short-term, intense activity. *Canadian Journal of Physiology and Pharmacology* 70: 157–165.

Spriet LL (1992). Anaerobic metabolism during exercise. In: *Exercise metabolism* (ed. Hargreaves M and Spriet L). 2nd edn. Champaign, IL: Human Kinetics, pp. 7–27.

Walker JB (1979). Creatine: biosynthesis, regulation and function. *Advances in Enzymology and Related Areas of Molecular Medicine* 50: 177–242.

Williams MH *et al.* (1999). *Creatine: the power supplement*. Champaign, IL: Human Kinetics.

Notes

1. The oxygen deficit represents the additional oxygen uptake that would be required to provide energy for muscular work at a work rate above 100% VO_{2max}.

2. The reader may find the following definitions useful:
   Maximal exercise is used by different authors to mean:
   (1) exercise at an intensity that elicits maximal oxygen uptake (VO_{2max})
   (2) completing a given amount of work or distance in the fastest possible time, or
   (3) all-out effort as in sprinting or a maximal voluntary isometric contraction. This is the definition used in this chapter.

   Submaximal exercise usually refers to exercise at an intensity less than that eliciting 100% VO_{2max}. Submaximal *steady-state* exercise usually refers to an exercise intensity that would elicit less than 80% VO_{2max}.
   Supramaximal exercise refers to exercise at an intensity above that eliciting 100% VO_{2max}.

3. A kinase reaction is an enzyme-catalysed reaction involving the transfer of an inorganic phosphate group ($H_2PO_4^{2-}$ usually abbreviated as P_i).

Middle-distance events

Learning objectives

After studying this chapter, you should be able to …

1. describe the oxygen cost of middle-distance running;

2. describe the relative contributions to energy metabolism from phosphocreatine breakdown, anaerobic glycolysis, and carbohydrate oxidation during middle-distance running;

3. give a general description of the glycolytic pathway;

4. describe the regulation of glycolysis;

5. describe how glucose is taken up by muscle from the blood;

6. explain the importance of lactate formation;

7. discuss mechanisms of fatigue in middle-distance events;

8. understand how increasing muscle buffer capacity can improve middle-distance running performance.

Introduction

> In middle-distance events the total amount of work done greatly exceeds that which is available from the phosphagen stores

Many sports involve intense exercise of a few minutes' duration, and some typical examples are shown in Table 4.1. In addition to these events, most of the combat sports, including boxing, wrestling, and judo, require the competitor to be able to perform multiple rounds of 3–5 min duration, with only a short recovery period between rounds. In all of these events, a high power output must be sustained for the duration of the event, and the total amount of work done greatly exceeds that which is available from the phosphagen stores (i.e. the alactic anaerobic processes described in Chapter 2). If phosphocreatine (PCr) were the only available fuel, the athlete would not be able to cover more than about 200–300 m before fatigue ensued, and there is no possibility of storing sufficient PCr in the muscles to last longer than this.

Oxidative metabolism makes the major contribution to energy production when the exercise duration exceeds about 1–2 min, but, at least for exercise intensities that can be sustained for less than about 10 min, the rate at which energy must be supplied to the working muscles exceeds the maximum rate of the oxidative processes. This chapter uses the example of the middle-distance track runner to describe the metabolic processes occurring and to consider the

Table 4.1 Current (at 1 March 2010) world records (min:s) in events where a high contribution from anaerobic glycolysis is required.

		Men	Women
Track cycling (individual 4000 m pursuit)		4:11.114	–
Track cycling (team 4000 m pursuit)		3:57.280	–
Rowing	2000 m single sculls	6:33.35	7:07.71
	2000 m pairs	6:14.27	6:53.80
Running	800 m	1:41.11	1:53.28
	1500 m	3:26.00	3:50.46
	5000 m	12:37.35	14:11.15
Swimming	400 m freestyle	3:40.07	3:59.15
	400 m individual medley	4:03.84	4:29.45

causes of fatigue and potential limitations to performance in events taking place over this time scale. The metabolic responses are closely related to the duration of the event, and we begin by considering the 1500-m event, which takes about 3.5 and 4 min for elite male and female runners, respectively.

Energy and oxygen cost of middle-distance running

> For every litre of oxygen consumed, about 21 kJ (5 kcal) of energy is expended when the predominant energy source is carbohydrate

The energy cost of running at different speeds can be measured on the treadmill. For convenience, the energy cost is normally expressed as the rate at which oxygen would be used if all of the energy requirement were met by oxidative metabolism. For every litre of oxygen consumed, we know that about 21 kJ (5 kcal) of energy is expended when the predominant energy source is carbohydrate. From results such as these, we can calculate that the total energy cost of running at world record speed for 1500 m is about 366 ml of oxygen per kg body mass, or 105 ml of oxygen per kg body mass per min if a constant speed is assumed. The oxygen consumption, however, increases only slowly and it takes about 2 min before it reaches its maximum value. Even when this maximum rate of oxygen consumption is reached, it will only be about 80–85 ml/kg/min for a world-class middle-distance runner (see Figure 4.1). Neither talent alone

Figure 4.1

The energy cost of running 1 mile (1604 m) in 4 min is about 0.2 kJ/kg/min, which corresponds to an oxygen cost of about 84 ml/kg/min, and not all of this energy can be met by aerobic metabolism. For a runner with a high aerobic capacity (70 ml/kg/min), about 80% of this energy can be met by oxidative metabolism, leaving 20% to be met by anaerobic metabolism.

nor the most intensive training will allow the athlete to achieve a value greater than this. Clearly, the shortfall in energy demand must be met by metabolic processes that do not involve oxygen. The line of oxygen consumption in Figure 4.1 demonstrates two important facts. First, if this were the only energy source, this would set the maximum running speed as shown on the right-hand axis, so the athlete would be able to accelerate only slowly, reaching maximum speed after about 700–800 m. Secondly, the maximum running speed would be closer to the 10 000-m speed than the 1500-m speed.

> The energy source for the sudden acceleration to running speed at the onset of a race is provided by the ATP and PCr stores in the muscle

When the gun goes to signal the start of the race, the athlete must increase the rate of energy supply from the resting level standing on the start line to something like 20–30 times this rate, and this clearly involves some major biochemical and physiological adjustments. The energy source for the sudden acceleration to running speed is provided by ATP and PCr stored in the muscle and therefore readily available to meet the immediate need. These mechanisms are described in detail in the preceding chapter. The high activity of creatine kinase, the enzyme that catalyses PCr hydrolysis, allows the muscle ATP concentration to be relatively well maintained at the expense of PCr, and the PCr concentration will fall rapidly. The muscle concentration of ADP, AMP, and inorganic phosphate (P_i) will begin to rise, signalling to the cell that the demand for ATP has exceeded the ATP resynthesis rate.

> Anaerobic glycolysis is important to ATP supply in the first few minutes of exercise

Because the amount of PCr available within the muscle cell is severely limited, the falling ATP and rising ADP, AMP, and P_i concentrations can be seen as signals of the need to supplement the energy available from the phosphagens. Because the oxidative processes are relatively sluggish, and also the maximum rate of energy supply from this source is limited, the muscles call on an alternative process involving the breakdown of carbohydrate stored within the muscles. The reactions of this pathway do not involve oxygen, so it is an anaerobic process, and because the fuel broken down to release energy is glycogen or glucose, the process is termed glycolysis (*gluco = glucose units; lysis = breakdown*). The rate at which anaerobic glycolysis can supply energy and the amount of ATP that can be made available in this way depend on a number of factors, including training status, muscle mass, and substrate availability. We can calculate, however, that the peak rate of glycolysis can supply ATP at somewhat less than half the maximum rate of PCr hydrolysis, but rather more than 1.5 times the maximum rate of aerobic energy supply (Table 4.2). The relative contribution of anaerobic

	Maximal rate of ATP resynthesis (mmol ATP/kgdm/s)	Delay time
Fat oxidation	1.0	>2 h
Glucose (from blood) oxidation	1.0	~90 min
Glycogen oxidation	2.8	Several minutes
Glycolysis	4.5	5–10 s
PCr breakdown	9.0	Instantaneous

Table 4.2 Maximal rates of ATP resynthesis from anaerobic and aerobic metabolism and approximate delay time before maximal rates are attained following onset of exercise.

metabolism to the total energy demand is about 60% in an event lasting about 2 min, falling to about 30% in a maximum effort of 4 min duration, and decreasing to 5% over 30 min. Even though oxidative metabolism is the main source of energy for events lasting more than about 2 min, these metabolic pathways are not discussed until Chapter 5: this chapter focuses on anaerobic glycolysis.

Glycolysis

Glycolysis, meaning the lysis or breakdown of glucose to pyruvate, does not use oxygen and takes place in the cytoplasm of the cell. It is a metabolic pathway that can generate ATP at a rapid rate and is a very important means of resynthesizing ATP in the first few minutes of exercise.

Fuels for glycolysis

Glucose and glycogen are the two main fuel sources for glycolysis

There are two main sources of fuel that can enter the glycolytic pathway: glucose and glycogen. Glucose is a sugar with six carbon atoms, and glycogen is a polymer of glucose, consisting of many thousands of glucose molecules linked together. As with all carbohydrates, these molecules are made up of atoms of carbon, hydrogen, and oxygen. As in water, the hydrogen and oxygen are present in a ratio of 2:1. The structure of the glucose molecule is shown in Figure 4.2, although this is only one of the two possible forms of the glucose molecule. Glucose has six carbon atoms, and is therefore classified as a hexose (hex = 6; -ose = sugar). Other important sugars are commonly either hexoses (examples are fructose and galactose) or pentoses (i.e. they have five carbon atoms: examples are ribose and

Figure 4.2

Structure of two monosaccharides (simple sugars), glucose and fructose, that can combine to form the disaccharide sucrose.

glucose
$C_6H_{12}O_6$

fructose
$C_6H_{12}O_6$

sucrose
$C_{12}H_{22}O_{11}$
(**O-β-D-fructofuranosyl-(2→1)-α-D-glucopyranoside**)

deoxyribose, and these important sugars are components of the nucleic acids, which are described again in Chapter 7; ribose also makes up a part of the ATP molecule). Glycogen has a complex structure and is in many ways similar to starch, which acts as a storage form of glucose in plants. One advantage of the polymer form is that it occupies much less space, but also, being almost insoluble, it can be stored without requiring large amounts of extra water to be retained by the cells. Even so, about 2–3 g of water is retained in the cell for each gram of glycogen that is stored.

The total amount of carbohydrate stored in the body is small

Glycogen is stored mainly in the muscles and in the liver: the liver glycogen store can be broken down to glucose and released into the bloodstream where it is available to all tissues to act as a fuel. This is especially important for the brain, which relies heavily on blood glucose as a fuel, and for other tissues such as the red blood cells, for which blood glucose is the only fuel that they can use. The muscle glycogen store has the advantage that it is more immediately available when the muscles are called on to do work, but it is not so easily available to other tissues. The total amount of carbohydrate stored in the body is small, with a maximum of about 100 g in the liver and 400–500 g in the muscles: these amounts depend on the preceding diet, as discussed later, and will be reduced by fasting and by exercise.

> Fats (other than the glycerol component of triglycerides) cannot be converted to carbohydrate

Most of the carbohydrate used by the body is derived from dietary carbohydrates, provided that a normal mixed diet is consumed. The dietary carbohydrates are a mixture of different types of sugars in the form of polymers (starch and glycogen), short-chain polymers (dextrins), disaccharides (e.g. sucrose, maltose, and lactose) and monosaccharides (e.g. glucose and fructose), but all must be converted to glucose before use by the body. Some of the carbon skeletons of the amino acids that make up dietary proteins can be converted by the liver to glucose, which can then be used immediately or stored as glycogen. There is, however, no mechanism for converting fats, whether these come from the diet or are in the form of stored fat in the body, to carbohydrate: only the glycerol component of stored fats can be salvaged and converted by the liver into glucose. Conversely, excess carbohydrate (and protein) in the diet can be converted readily into fat.

The glycolytic pathway

> Glycolysis is the breakdown of glucose to pyruvate and does not use oxygen

Glycolysis, as described earlier, does not use oxygen and takes place in the cytoplasm of the cell. The main reactions involved in the breakdown of carbohydrate by anaerobic metabolism are shown in Figure 4.3. This simplification has omitted a number of important steps that are described in more detail later in this chapter. The important points to note are:

- stored carbohydrate, in the form of six-carbon sugars, which may be glycogen or glucose, is the starting point;
- two molecules of the three-carbon pyruvate result as the end point;
- some of the chemical energy in the glucose molecule is conserved by conversion of ADP to ATP;
- nicotinamide adenine dinucleotide (NAD), which is involved as a cofactor in one of the reactions of glycolysis, is simultaneously converted to its reduced form, NADH.

Figure 4.3

Key steps in the breakdown of carbohydrate in the glycolytic pathway. Several intermediate steps are omitted for the sake of clarity, but these are included in Figure 4.4. Note that all the reactions below the level of fructose 1,6-diphosphate occur in duplicate as two molecules of the three-carbon 1,3-diphosphoglycerate are formed. Two ATP molecules are broken down to ADP to prime the pathway (if starting from glucose; only one ATP if starting from glycogen), and four ATPs are produced. The net gain is therefore only two (from glucose) or three (from glycogen) ATP molecules. See text for further details.

The significance of this last fact will become clear shortly, but first some of the key reactions in the glycolytic pathway should be considered.

 The enzyme glycogen phosphorylase splits off glucose units from glycogen

If muscle glycogen is the starting point, the first reaction involves the splitting off from the large glycogen molecule of a single glucose molecule, which is released as glucose 1-phosphate, and this in turn is rapidly converted to glucose 6-phosphate. The first step is catalysed by the enzyme glycogen phosphorylase (a lysing, or splitting, reaction involving a phosphate group: in this case, it is the glycogen molecule that is split), and this is a key step in the regulation of glycogen metabolism. It is important to note that the phosphate group that is added to the glucose molecule comes from inorganic phosphate that is present within the cell. The addition of the phosphate group is important as it primes the glucose molecule for the subsequent reactions, and it also stops it escaping from muscle cells: glucose can cross the cell membrane in both directions, but the addition of a phosphate group to the glucose molecule ensures that this valuable fuel is not lost from the cell.

 Glucose from the blood is transported into muscle fibres by the transporter GLUT4

Glucose present in the blood can be taken up by muscle cells, particularly in active muscle, and used as the starting point for glycolysis. This requires a specific transport process to convey the glucose molecule across the cell membrane, and this transporter (called GLUT4) is closely linked to the enzyme hexokinase (kinase = adding a phosphate group, to hexose, a six-carbon sugar), which ensures that glucose entering the cell has a phosphate group attached to it and is thus effectively trapped. In this case, the phosphate group is added to carbon 6 of the glucose molecule to form glucose 6-phosphate. There is another important difference in that energy has to be added to the system to drive the hexokinase reaction, and as usual, the source of this energy is the breakdown of ATP to ADP. At a time when the need for ATP synthesis is high, it might seem strange that the cell must use some of its limited resources to begin the process of producing energy from the breakdown of glucose. At times when the energy demand is high, however, the glycogen stored within the cell will be the main fuel source and this does not require the investment of an ATP molecule to initiate the process.

The enzyme phosphofructokinase catalyses a key regulatory step in glycolysis

Glucose 6-phophate is then converted to fructose 6-phosphate. As both glucose and fructose are six-carbon sugars, this process simply involves a rearrangement of the carbon atoms, converting one hexose sugar into the other. The next reaction in the glycolytic pathway is a key regulatory step and results in the formation of fructose 1,6-diphosphate (FDP). As its name indicates, FDP is a molecule of fructose with two phosphate groups attached: the first of these is the one bonded to glucose during the initial activation process and the second is added from a molecule of ATP. So far, then, two molecules of ATP have been broken down if the starting point is glucose, and one if glycogen is the starting point. The reaction responsible for the formation of FDP is catalysed by the enzyme phosphofructokinase. Again, the name describes what happens: a phosphate group is added (a kinase reaction) to fructose 6-phosphate. This reaction is important as it is one of the steps that regulates the rate at which the whole pathway proceeds, and the regulatory mechanisms are described later in more detail.

The net gain from anaerobic glycolysis is two ATP molecules starting from glucose, or three if glycogen is the fuel used

The six-carbon molecule FDP is then split into two three-carbon molecules, each with one phosphate group attached. A number of sequential steps follow, in which a second phosphate group is added before both phosphate groups are removed from the three-carbon molecules and attached to ADP, resulting in resynthesis of two molecules of ATP. Because there are two three-carbon molecules undergoing this series of reactions (once for each of the three-carbon molecules) and each results in the formation of two ATP molecules, the overall result is the formation of four ATP molecules. We must remember, however, the initial investment of one ATP when the starting point was glycogen, and two ATPs starting from glucose. The net effect is therefore a gain of two ATPs starting from glucose and three ATPs if, as is usual when the exercise intensity is very high, glycogen is the fuel used. The end-product is the three-carbon molecule pyruvate, with two molecules of pyruvate being produced from each six-carbon starting point.

For the sake of completeness, and to show the points at which ATP is consumed or formed, the complete series of reactions comprising the glycolytic pathway is shown in Figure 4.4. Each of these reactions is important, but the key steps involving ATP formation are those outlined above. There is, however, one more reaction that is of great significance to the cell. This is the conversion of glyceraldehyde 3-phosphate to 1,3-diphosphoglyceric acid. In this reaction, a phosphate group is added and two hydrogen atoms are removed: one of the hydrogen atoms is attached to the cofactor NAD, converting it to NADH, and the other is released as a free hydrogen ion. Both of these hydrogen ions have vital consequences for the cell: there is a very limited amount of NAD available within cells, and the formation of NADH changes the potential for many different reactions in the cell, and the free hydrogen ion released causes the local environment to become more acid. Conversion of NADH back to NAD requires further reactions to take place if energy production by glycolysis is to make a meaningful contribution to energy production.

Regulation of glycolysis

Various control mechanisms exist to ensure that glycolysis proceeds at the required rate

The rate of energy supply within each active muscle cell must be matched exactly to the energy demand, and this requires a precise control mechanism to ensure that glycolysis proceeds at the required rate. It is also important to ensure that the body's fuel stores are used as efficiently and economically as possible. Many different factors can affect the rate of the key enzymes involved in the glycolytic pathway, and the first important control point involves the enzyme glycogen phosphorylase. This enzyme exists in two forms, one of which, designated phosphorylase a, has a higher activity than the other, phosphorylase b. The hormone adrenaline, which is secreted at times of stress, such as before a race, promotes the conversion of phosphorylase b to phosphorylase a. Phosphorylase a, however, can only begin to break down glycogen in muscle if the concentration of calcium is above a certain threshold level, and this concentration is reached only when a nerve impulse arrives at the muscle cell and stimulates it to contract. In this way, breakdown of glycogen begins only when the cell is activated, thus preventing it from being used by the cells at times when the demand for energy is not high, but also ensuring that the rate of energy production by anaerobic glycolysis is increased as soon as the energy demand is increased.

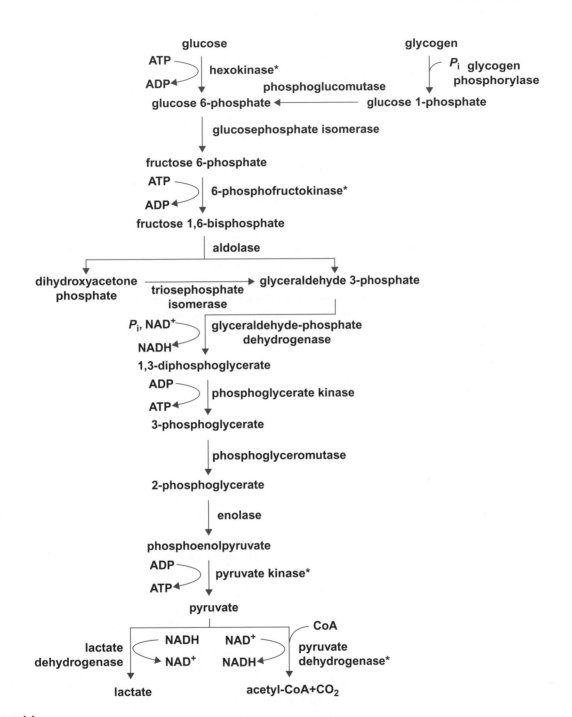

Figure 4.4

The glycolytic pathway in detail.

The entry of blood glucose into the pathway is regulated by the availability of the GLUT4 glucose transporters to carry the glucose across the cell membrane and by the activity of the enzyme hexokinase, which primes the glucose molecule by the addition of a phosphate group. In high-intensity exercise, glycogen stored within the muscle supplies most of the substrate for the glycolytic pathway, and blood glucose makes only a minor contribution. Even though the hexokinase enzyme is activated when the concentration of phosphate inside the cell rises, indicating a shortfall in energy supply, the enzyme is also inhibited by glucose 6-phosphate (G6P). When glycogen breakdown is occurring rapidly, the concentration of G6P within the cell rises (see Figure 4.4), thus slowing the entry of blood glucose into the pathway.

Other regulatory mechanisms function by responding to changes in the energy status of the cell: if the ATP concentration is high, the rate of glycolysis is low, and this is important in conserving the limited carbohydrate stores at a time when the muscle can use fat as a fuel. As the ATP concentration begins to fall and the ADP and inorganic phosphate concentrations begin to increase, this acts as a signal to the muscle that aerobic mechanisms are failing to keep pace with the energy demand, and the rate of glycolysis is increased to maintain the intracellular ATP level.

The reaction catalysed by phosphofructokinase is the rate-limiting reaction in the glycolytic pathway

The reaction catalysed by phosphofructokinase (PFK) is an important regulatory step, and this enzyme is affected by the intracellular concentrations of a number of components: in particular, its activity is inhibited when ATP levels are high (as happens at rest or during low-intensity exercise when the energy demand is met largely by fat oxidation) and stimulated when ADP and AMP levels are high (as happens when the ATP concentration falls, signalling an inability of oxidative metabolism of fat to meet the energy demand). PFK is inhibited by hydrogen ions, and the increased level of acidity in the muscle that occurs when glycolytic rates are high is often cited as a cause of fatigue, acting in effect as a safety mechanism to prevent pH from falling to levels that might cause irreversible damage to the cell. However, the potential inhibition due to falling pH is generally overcome by the simultaneous fall in the ATP concentration and the increases in ADP, AMP, and ammonia concentrations that will act to increase its activity. This is discussed further below. The activity of PFK is also inhibited when the concentration of citrate is high: citrate concentrations are high when oxidative metabolism is proceeding at high rates, and although there is some dispute as to how important this mechanism is in human skeletal muscle under normal conditions, this is another possible way of integrating the rates at which oxidative metabolism and anaerobic glycolysis occur. Integration is important to ensure that the energy demand is met whenever possible, but also that the demand is met without using more of the limited carbohydrate store than is necessary. There is never likely to be a shortage of fat in the body, so even though there is no mechanism for obtaining energy from fat without involving oxygen, it makes sense to use fat as a fuel whenever this is possible.

Leaders in the field

Eric Hultman was responsible for pioneering studies with fellow researcher Jonas Bergstrom at the Karolinska Institute of Stockholm in the 1960s. These led to the development and use of Bergstrom's muscle biopsy needle to remove small samples of muscle (and liver) from the human body. The technique, now in use throughout the world, allowed the metabolic changes in the muscle of humans to be measured for the first time, providing significant insight into the mechanisms regulating muscle metabolism, fatigue, and recovery. The early studies also gave rise to the carbohydrate loading diet. The first studies on which this was based were published in 1966 and 1967, but it was only in 1969 that this method of enhancing performance was first used in the world of sport. Since then, the importance of carbohydrate for endurance performance has been universally recognised. In the late 1980s he was largely responsible for the early studies on the effects of creatine supplementation on muscle metabolism and the performance of high-intensity effort.

Regeneration of NAD

Conversion of pyruvate to lactate allows the regeneration of the NAD that was converted to NADH earlier in the glycolytic pathway

The amount of NAD within the cell is extremely small, and very soon it would all be converted to NADH if glycolysis were to proceed rapidly without some mechanism for regeneration of NAD. Although the NAD molecule has a complex structure, it can be thought of simply as a recipient or donor of hydrogen atoms in biochemical reactions where these hydrogen atoms must be removed from or added to other molecules. In the glycolytic pathway, this occurs in the reaction catalysed by glyceraldehyde-3-phosphate dehydrogenase, as shown in Figure 4.4: as its name implies, this dehydrogenase reaction involves the removal of hydrogen atoms (two of them) from the substrate, which is glyceraldehyde 3-phosphate (G3P). In the process, a phosphate group is added and the end-product of the reaction is 1,3-diphosphoglyceric acid, so this is clearly a complex reaction requiring that the substrates (G3P, NAD, and P_i) be brought together so that they can interact. It is even more complex, in that NAD is a positively charged molecule (NAD^+), and the addition of a hydrogen ion (H^+), which also carries a positive charge, is balanced by the addition of two negatively charged electrons so that the end-product (NADH) has no net electrical charge.

When the demand for energy is low, NADH is normally reconverted back to NAD by transfer of the unwanted hydrogen atom to oxygen with the end-product being water. The reactions that allow this to occur take place in the mitochondria and are described in the next chapter. However, during intense exercise, the rate of glycolysis and therefore the rate at which NAD is converted to NADH exceed the maximum rate at which the oxidative system can regenerate NAD, so clearly another mechanism must be available.

When the rate of glycolysis exceeds the rate at which NAD can be regenerated by oxidative metabolism, the cell makes use of the large amount of pyruvate formed as the end-product of the glycolytic sequence of reactions. Conversion of pyruvate to lactate involves the conversion of NADH to NAD. This allows the regeneration of the NAD that is consumed earlier in the glycolytic sequence. The lactate that is formed can leave the cell (as indeed the pyruvate can do): the difference is that lactate accumulates in much higher concentrations. If the pyruvate did leave the cell, it would not be available for regeneration of NAD.

Oxidative metabolism of carbohydrate

The complete oxidation of the pyruvate allows much more of the energy stored in the glucose molecule to be made available in the form of ATP

Regeneration of NAD by conversion of pyruvate to lactate can be seen as only a short-term solution to the problem, as the amount of lactate (or more correctly, the amount of hydrogen ions that accompany lactate accumulation) that the body can tolerate is severely limited. The alternative is complete oxidation of the pyruvate to carbon dioxide and water: this allows not only regeneration of the NAD, allowing glycolysis to continue, but also much more of the energy stored in the glucose molecule to be made available to the cell in the form of ATP. The complete conversion of the three-carbon pyruvate molecule to CO_2 and H_2O requires transfer of the pyruvate into the mitochondria where the necessary enzymes and cofactors are located.

The steps that involve transfer of the hydrogen atoms from NADH to oxygen, resulting in the regeneration of NAD and making energy available for the resynthesis of ATP, are part of the electron-transfer chain. In this, the two electrons and the hydrogen ion transferred to NADH follow a number of sequential reactions in which sufficient energy is made available at three points for a phosphate group to be bonded to ADP for each NADH formed. All of these reactions also take place within the mitochondria.

The first step in the mitochondrion is the conversion of pyruvate to acetyl-CoA

The first step in this process involves the entry of pyruvate molecules into the mitochondria where the enzymes of the oxidative pathway are located, and this

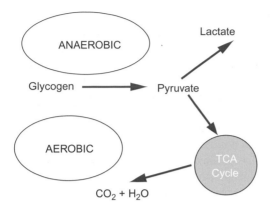

Figure 4.5

Reactions involved in the further metabolism of pyruvate. Whether it is converted to lactate or to acetyl-CoA depends on a number of factors, including the availability of oxygen in the cell and the rate of glycolysis (resulting in pyruvate formation) relative to the rate of pyruvate removal by uptake into the mitochondria.

is clearly facilitated when the muscle has a high content of mitochondria. The first step in the further metabolism of pyruvate involves the reactions shown in Figure 4.5. These reactions are catalysed by the enzyme pyruvate dehydrogenase (PDH). As the name implies, this enzyme catalyses the removal of hydrogen atoms from the pyruvate molecule (with NAD again being the acceptor), but it also does much more and in fact is a complex of enzymes that perform different functions. It may seem odd that the first stage of the process that is designed to restore the level of NAD in the cell in fact consumes this valuable compound, but this can be looked on as an investment of the cell's resources that will pay an immediate (or at least almost immediate) dividend. It must also be remembered that the NAD used here comes from within the mitochondria, whereas glycolysis has used NAD from the cell cytoplasm. To prime the pyruvate molecule for the subsequent reactions, it is attached to a molecule of coenzyme A (CoA) and, at the same time, one of the carbon atoms is lost as carbon dioxide, yielding acetyl-CoA. Acetyl-CoA then enters the tricarboxylic acid (TCA) cycle (also known as Krebs cycle after Hans Krebs, the scientist who first described it). This sequence of reactions and the events that follow in the electron-transport chain are described in detail in Chapter 5.

Fatigue mechanisms in middle-distance events

The maximum accumulation of lactate within the muscle occurs at the end of exercise that causes exhaustion in about 3–7 min

Fatigue can be defined as the inability of a muscle to maintain a required rate of work. Although the subjective sensation of fatigue is familiar to everyone, the underlying causes are not clearly understood at the present time. In maximal work lasting about 1–2 min, the PCr content of the working muscle falls almost to zero and the ATP content falls by about 40%. Energy supply from lactate formation and from oxidative metabolism is still possible, but these systems alone are not able to generate sufficient power to sustain work of very high intensity. Measurements on elite-level sprinters show that maximum speed is reached after about 3–4 s (at about 30–40 m of running) and declines thereafter. As the distance increases, so the maximum speed that can be sustained also falls, reflecting the gradual decrease in the peak rate of energy supply: Figure 4.6 shows the world best performance for increasing distances up to 10 000 m. This can be taken to represent the line of maximum energy supply rate (allowing for some slight decreases in the efficiency of running as the speed increases).

Once the PCr store has been depleted, it can clearly make no further contribution to energy supply. No further fall in the muscle ATP content occurs after the first few seconds, so it can be assumed that the available ATP reserve has also been exhausted. In this situation, it is possible, and indeed even probable, that the inability to continue working at a high rate is the result of depletion of the phosphagens. For the 1500-m runner, the initial speed is much less than that of the sprinter, and the phosphagen stores can last longer, but the PCr will be almost completely depleted by about the half-way point, and cannot make much contribution to energy production in the second half of the race. Any remaining PCr is likely to be used during the final sprint if an all-out effort is required in the closing stages of the race.

In high-intensity exercise, some of the lactate formed by anaerobic glycolysis accumulates within the muscles where it is produced and some diffuses out of the cells and reaches the blood. Although measurements of blood lactate are of

Figure 4.6

Running speed at the world records for middle-distance running for men and women. This represents the maximum possible rate of energy supply over these distances.

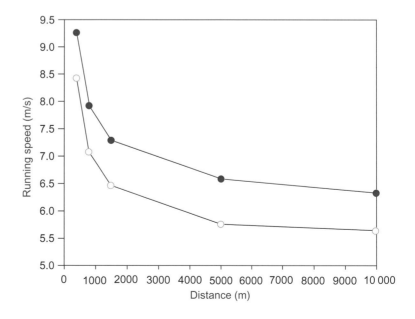

interest and give some indication of what is happening within the muscle, muscle lactate itself is of far greater importance—fatigue, after all, occurs within the muscle, not in the blood. Measurements of muscle lactate content after exercise have shown that the maximum accumulation of lactate within the muscle occurs at the end of exercise that causes exhaustion in about 3–7 min. Lower muscle lactate concentrations are seen after exhausting work lasting less than 2 min or more than 10 min. This suggests that if lactate accumulation is the cause of the fatigue experienced by the muscle, this may be true only for work lasting within that narrow range of about 3–7 min. There is, however, no evidence to support the suggestion that lactate itself is responsible for fatigue despite the close relationship between the concentration of lactate in muscle and blood and the subjective sensation of fatigue.

> The pH in resting muscles is about 7.0, and can fall to 6.3 during high-intensity exercise

An inevitable consequence of energy production by anaerobic glycolysis is the fact that, along with the lactate formed, hydrogen ions are also produced, causing the internal environment of the cell to become more acidic. The acidity is normally expressed as pH; the lower the pH of a solution, the more acid it is. The pH scale is such that a value of 7.0 represents a neutral solution—one that is neither acidic nor alkaline: values less than 7.0 indicate an acidic solution, and values greater than 7.0 an alkaline one. The pH in resting muscles is about 7.0, and it is important for most cellular processes that the pH is kept within a narrow range. Most enzymes, for example, function optimally only within a specific range of pH. For this reason, a number of buffers—substances that can absorb or release hydrogen ions without any change in the acidity level—are present in the cells and in the extracellular space.

> Hydrogen-ion accumulation in the cell may cause fatigue by inhibiting the reactions of anaerobic glycolysis or interfering directly with the contractile mechanism

As anaerobic glycolysis proceeds, the hydrogen ions produced along with the lactate overcome the buffering capacity of the cell and cause the pH to fall—the cell becomes more acid. At the point of fatigue the pH may fall as low as 6.3. The blood pH is normally slightly alkaline at rest, about 7.4, and falls after exhausting high-intensity exercise to about pH 7.0 or even slightly less, indicating a release of hydrogen ions from the working muscles. This helps to increase the amount of buffer available to counteract the acidifying effect within the muscles. The changes in the acidity of the muscle have important consequences for muscle function. As the pH of the muscle cell falls, the concentration of free hydrogen ions increases. High levels of hydrogen ions in the cell may cause fatigue in two ways; first, they

Figure 4.7

Increasing levels of acidity within the muscle cell causes a subjective sensation of pain and fatigue, as well as decreasing exercise performance. The subjective sensations arise from stimulation of free nerve endings by increasing hydrogen ion concentration. Decreased muscle performance arises from a number of mechanisms, as described in the text.

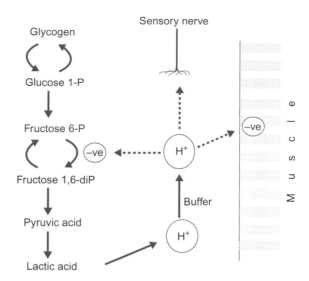

interfere with the series of chemical reactions responsible for energy production by anaerobic glycolysis, thus decreasing the capacity of the muscle to produce energy in this way, and secondly, they may interfere directly with the contractile mechanism itself (as illustrated in Figure 4.7). Both of these effects result in fatigue, the first by reducing the energy available to the cell and the second by preventing the cells from using the available energy to perform work.

However, although it is easy to use purified enzymes in a test tube to demonstrate that the maximum activity of phosphofructokinase, one of the key enzymes in the glycolytic pathway, is reduced when the environment becomes more acid, there has been some debate about the importance of the inhibitory effect of increasing acidity in the muscle cell when other regulatory mechanisms are acting to stimulate this enzyme. The increasing concentration of hydrogen ions also stimulates free nerve endings in the muscle, giving rise to the characteristic painful sensations that accompany high-intensity exercise. We may be unsure of the exact mechanism, but fatigue almost certainly results from the fall in pH, which is an inevitable consequence of energy provision by anaerobic glycolysis with subsequent lactate formation, although the lactate itself is not responsible.

In exercise of longer duration the pH of the muscle at the point of exhaustion is higher than is the case in exhausting work of 3–7 min duration, while the muscle PCr content is similar or perhaps even higher. This time scale includes most of the events in Table 4.1. Another cause for the fatigue experienced in this type of work must be sought, but at present the mechanism responsible is not clear. In exercise of long duration, greater than about an hour, depletion of the muscle glycogen stores may be the cause of fatigue, but this does not seem to be true for shorter-duration efforts. However, performance is likely to be reduced if the exercise begins with an inadequate store of glycogen in the muscles, and performance may be improved by manipulation of the diet to increase the pre-exercise muscle glycogen store. This is discussed further in Chapter 5.

Recovery after exercise

The key components of the recovery process are the restoration of the muscle PCr and ATP levels, removal of accumulated lactate, restoration of the normal pH, and recovery of the muscle glycogen stores

The middle-distance runner who has run as fast as possible crosses the finishing line completely exhausted. Recovery is important where another round of competition follows, and is also important in training where repeated high-intensity efforts are made with only a short rest interval. The recovery process must involve reversal of the changes occurring during the fast run and restoration of the capacity for high-speed running. The key components of the recovery process must be restoration of the muscle PCr and ATP levels, removal of the lactate that has accumulated, restoration of the normal pH level, and recovery of the muscle glycogen stores. All of these processes must be completed before full recovery occurs, and the energy necessary for recovery must ultimately be derived from the oxidative metabolism of the body's fuel stores or from food eaten during the recovery period. The pathways involved in making ATP available by oxidation of these fuels are described in detail in the next chapter: for the moment, we assume that sufficient ATP is readily available to the muscle cell.

Recovery of the ATP and PCr stores occurs within a few minutes

Replenishment of the PCr store within the muscles involves a simple reversal of the creatine kinase reaction: the total size of this store is fairly small and resynthesis of PCr occurs fairly quickly (although not as rapidly as the forward reaction involving breakdown of PCr). Complete restoration of the resting PCr levels takes about 5–10 min, but the process is an exponential one, and is more than 50% complete within the first 30–60 s of recovery. This process has already been described in more detail in the preceding chapter.

Getting rid of the lactate and restoring the pre-exercise pH takes much longer

Removal of lactate and restoration of the pre-exercise pH is a much slower process, and depends on a number of factors, including the muscle activity level during the recovery period. If the athlete sits down or lies down after a maximum effort, lactate and hydrogen ions diffuse out of the muscle cells, where the concentration is high, into the blood, where the concentration is lower. Lactate and hydrogen ions move out of the muscle mostly via facilitated diffusion using transport proteins in the sarcolemma that are called monocarboxylate transporters (MCT). The isoforms of these present in skeletal muscle are MCT1 and MCT4 and there is some evidence that the expression of these can be increased following a period of high-intensity interval training. After leaving the muscle and entering the circulation, some of the lactate is then taken up by the liver, where it can be converted back to glucose (muscle does not possess the enzymes that are necessary to synthesize glucose from lactate), or it can go to active tissues (especially to the heart, which is always active). Liver, skeletal muscle, and heart muscle all contain a high activity of the enzyme lactate dehydrogenase, which catalyses the reversible interconversion of lactate and pyruvate. During recovery lactate can be converted back to pyruvate. Active tissues, such as the heart, use it as a fuel to produce energy by oxidative metabolism, resulting in the conversion of the lactate to carbon dioxide and water, as described in Chapter 5. The hydrogen ions entering the blood react with bicarbonate ions there to form carbon dioxide and water by the following reaction:

$$H^+ + HCO_3^- \rightarrow H_2CO_3 \rightarrow H_2O + CO_2.$$

This reaction limits the fall in the pH of the blood that would otherwise occur. The carbon dioxide that is formed by this reaction is lost through the lungs. In the later stages of recovery, the blood bicarbonate stores are replenished by reducing the amount of carbon dioxide that is lost in the expired air, but restoration of the blood bicarbonate level can take an hour or more. It takes about the same length of time for the blood and muscle lactate concentrations to return to resting levels after an all-out effort, which means that a maximum effort cannot be repeated within this time.

Doing some activity speeds up the recovery process

The recovery process can be hastened if, instead of collapsing at the side of the track, the athlete undertakes an active recovery involving walking or slow running. This changes the fate of the lactate. Instead of mostly being taken up and used by the liver, it is now used as a fuel for producing energy by oxidative metabolism in the muscles that are active (particularly the Type-I fibres): these reactions are described in the next chapter. Even fairly strenuous running is effective in increasing the rate of lactate oxidation, especially in the trained muscles of runners. However, although this has the advantage of increasing the rate at which lactate is removed from the system, it does reduce the amount that is available for another important part of the recovery process, namely the replacement of the muscle glycogen stores.

Restoring the muscle glycogen used in exercise is likely to take at least 24 h

Replenishment of the muscle glycogen stores is the part of the recovery process that takes the longest time. Glycogen stored in the muscle can be broken down very rapidly: more than 100 g of glycogen can be converted to lactate during a race over 800 or 1500 m, and this represents about 25% of the total body carbohydrate store. Replacement of the muscle glycogen store requires a supply of glucose from the blood: some of this can come from the liver, which is busily converting lactate back to glucose, but most will come from the diet, and recovery can be accelerated if carbohydrate-containing foods are eaten at this time. Even so, it is likely to take at least 24 h before the muscle glycogen level is back to normal.

The slowness of these reactions involved in the recovery process explains why the recovery period after middle-distance running takes so long. If several rounds of competition are scheduled to take place in a single day, it is unlikely that the athlete will be able to produce their best performance in each race. The athlete who wins the final may not be the one who would have won if only a single race was involved, or may be the one who had easier races in the early rounds. This also shows the importance, where another race follows, of taking steps to ensure that recovery is as rapid and as complete as possible.

> ### Leaders in the field
>
> **Archibald Vivian Hill** (universally referred to as AV Hill) was born in 1886, and studied mathematics at Cambridge University. He moved to physiology and started research work on the nature of muscular contraction. He was influenced by the work of Fletcher and Hopkins on the problem of lactic acid in muscle, particularly in relation to the effect of oxygen upon its removal in recovery. This led him to study the dependence of heat production on the length of muscle fibre. In 1919 he began a long-term collaboration with Otto Meyerhof. Hill (who was a better than average 800-m runner) then began to apply the results obtained on isolated muscles to the study of exercise in man. After a long and distinguished career, Hill died in 1977.

Nutritional effects on the performance of the middle-distance athlete

Carbohydrate

Muscle glycogen content has to be at normal levels or higher if optimum performance is to be achieved

As with the sprinter, much of the training programme of the middle-distance athlete consists of repeated bouts of high-intensity efforts with variable rest periods. There is therefore a need to ensure an adequate dietary carbohydrate intake during periods of intensive training to maintain the muscle glycogen stores. In competition, glycogen availability is not usually a limiting factor, unless there has been inadequate recovery from the last exercise session and an inadequate carbohydrate intake. It does seem important, however, that the muscle glycogen content is at normal levels or higher if optimum performance is to be achieved. There is good evidence that performance is impaired if events begin with a low muscle glycogen content, and a few days of reduced training combined with a high carbohydrate diet should ensure an adequate muscle glycogen store before racing.

> Carbohydrate oxidation generates more energy per litre of oxygen used than fat oxidation

Most of the energy supply during exercise lasting a few minutes comes from degradation of carbohydrate, including both liver and muscle glycogen. Using carbohydrate as a fuel for oxidative metabolism has some advantages as the oxidation of carbohydrate generates 21.0 kJ (5.01 kcal) of energy for each litre of oxygen, while oxidation of fat makes only 19.6 kJ (4.68 kcal) available. This difference may seem small, but in situations where the capacity of the cardiovascular system to supply the working muscles with oxygen is a primary limitation to performance, this difference may be crucial. Consuming a high carbohydrate diet in the hours and days prior to competition suppresses fat use by the muscle and ensures the greatest possible use of carbohydrate.

In studies of high-intensity cycling, it has been shown that performance is impaired if muscle glycogen stores are depleted before exercise and improved if a high carbohydrate diet is fed for the last few days before competition (Table 4.3). Ingestion of a low carbohydrate diet is associated with a metabolic acidosis and this has potentially negative effects on performance. If an incremental exercise test is performed after these different diets and the blood lactate concentration is measured, the lactate–workload curve is shifted to the right after the low carbohydrate diet and to the left after the high carbohydrate diet (Figure 4.8). A rightward shift of this curve is normally an indication of improved fitness and is

Table 4.3 Exercise time to fatigue on a cycle ergometer at 100% of VO_{2max} after eating diets with different carbohydrate (CHO) contents.

	CHO content (%)	Exercise time (min)
Normal diet	43	4.87
Low-CHO diet	3	3.32
High-CHO diet	84	6.65

associated with a better performance. Here, however, the situation is reversed, and there is a loss of the normal association between lactate and fatigue. The reduced lactate formation after the low carbohydrate diet means that less energy is available from anaerobic glycolysis.

Techniques

Blood sampling. Blood is perhaps the most frequently sampled tissue in human investigations. The site and method of sampling can greatly influence the results obtained, so investigators must give careful thought to what they are measuring and how this is influenced by the method of sampling. The oxygen content of arterial blood, for example, is obviously very different from that of the venous blood that drains the muscles during exercise. The muscles, however, also remove various substrates, hormones and other compounds, and in turn add a range of metabolites and other products. Sampling from both the arterial blood and from a vein draining a muscle can give some information on what reactions are taking place in the muscle and can establish if net uptake or release is taking place. If some measure of blood flow can also be obtained, quantitative estimates of metabolic activities and rates of uptake or release of substances can be obtained.

Increasing buffer capacity

Ingestion of bicarbonate, carnosine, or β-alanine prior to exercise can improve performance in middle-distance events

Associated with the high rates of anaerobic glycolysis in middle-distance events is a profound fall in the intracellular pH of the muscle fibres. The development of intracellular acidosis in the working muscles is commonly cited as the cause of muscle fatigue during high-intensity exercise. It is well established that intracellular acidification can reduce the sensitivity of the contractile apparatus to Ca^{2+} and can also reduce the maximum Ca^{2+}-activated force that is generated. However, producing lactic acid is not a bad thing, as many coaches would suggest. High rates of anaerobic glycolysis allow high rates of ATP resynthesis and thus high power outputs. Acidosis can be avoided very simply by just running slowly, but that does not win any races. In any case, there is some evidence that acidosis helps to preserve muscle function during intense activity. As the muscle is activated, membrane permeability increases: sodium enters the cells and potassium leaves, leading to accumulation of potassium in the extracellular space and in the network of tubules that make up the T system and allow inward conduction of the action potential. This in turn results in inactivation of the membrane sodium

Figure 4.8

Relationship between blood lactate concentration and power output (expressed as a fraction of VO_{2max}) in incremental cycle ergometer tests carried out after a normal diet, a low carbohydrate (CHO) diet and a high carbohydrate diet. The low CHO diet results in a right shift in the lactate-workload curve, while the high CHO diet results in a left shift in the curve.

Figure 4.8

Relationship between blood lactate concentration and power output (expressed as a fraction of VO_{2max}) in incremental cycle ergometer tests carried out after a normal diet, a low carbohydrate (CHO) diet and a high carbohydrate diet. The low CHO diet results in a right shift in the lactate-workload curve, while the high CHO diet results in a left shift in the curve.

channels that generate and propagate action potentials, leading to loss of force production. Acidification of the muscle reduces the loss of muscle excitability and force generating capacity seen in depolarized muscles. The underlying mechanism is not clear but seems to involve a change in the membrane permeability to chloride, so that action potentials can still be propagated along the T system despite muscle depolarization.

In spite of these possible protective effects, there remain some negative consequences of high intracellular hydrogen ion concentrations, and it has been suggested that the ingestion of alkaline salts prior to exercise may improve performance in middle-distance events. This suggestion is supported by experimental evidence showing that administration of bicarbonate prior to exercise can improve performance in races over 800 or 1500 m. In the same way that bicarbonate neutralizes the excess stomach acid that causes indigestion, it can neutralize some of the excess acidity in the muscles. Although very little of the ingested bicarbonate enters the muscle cells, the higher extracellular pH and buffering capacity allow hydrogen ions to leave the exercising muscle at a faster rate. This allows more hydrogen ions, and more lactate, to be produced before the acidity within the muscle cell reaches a limiting level. The results of two different experiments that have demonstrated improved performance in races over these distances after the subjects consumed sodium bicarbonate before running are shown in Table 4.4.

	Control	Placebo	Bicarbonate
800 m	2:05.8	2:05.1	2:02.9
1500 m	4:18.0	4:15.6	4:13.9

Table 4.4 Effects of bicarbonate administration on 800-m and 1500-m racing performance (min:s). These results are taken from two unrelated studies that have produced very similar results.

Note that bicarbonate loading in this way does not directly alter the pre-exercise muscle pH; rather, the increased extracellular pH and buffering capacity allow a more rapid removal of H^+ (and lactate) ions from the muscle during exercise, so that it takes longer before the intramuscular H^+ ion concentration accumulates to a critical level. This is achieved by ingestion of a dose of about 0.3 g $NaHCO_3$ per kg body mass over the space of a few hours prior to the event, so for a 70-kg individual, the total dosage would be 21 g—about six level teaspoons. Sodium bicarbonate can be purchased as baking soda, a white powder that is readily available in most grocery stores. The powder can be placed in gelatin capsules (these can be bought from the chemists) and swallowed together with about one litre of water or cordial 1 to 2 h prior to exercise. Of course, the interaction of the bicarbonate with the acid environment of the stomach results in the evolution of carbon dioxide, and may be associated with unpleasant subjective symptoms. Hence, there is some risk of gastrointestinal distress (vomiting and diarrhoea) with this amount of $NaHCO_3$ intake, but these effects are probably less common and less severe than is commonly supposed. If they can be avoided, an improved performance is possible in those events where a metabolic acidosis is the main factor limiting performance. Obviously, any athlete considering bicarbonate supplementation should first try it out a few times in training to see if they can tolerate it. It would not be a good idea to take bicarbonate for the first time just before an important event.

Doses of less than 0.1g per kg body mass are not likely to be effective, but attempting to consume larger doses than 0.3 g per kg (which corresponds to about 20 g of $NaHCO_3$ for a 70-kg individual) is not likely to further improve performance and the risk of gastrointestinal upset will be increased. In other words, appropriate dosage is very important, and an accurate set of scales should be used when weighing out the sodium bicarbonate. There is some evidence that ingesting the sodium salts of organic acids, such as citrate, can also produce a similar performance-enhancing effect with less risk of gastrointestinal discomfort.

In general, sodium bicarbonate ingestion is safe when taken in the doses recommended above. Some people experience gastrointestinal distress such as bloating, flatulence, nausea, and diarrhoea. Excessive doses can cause severe alkalosis, which can result in muscle spasms and heart arrhythmias. Sodium bicarbonate supplementation is currently legal; whether its use is ethical or not is debatable.

Figure 4.9

Structure of the dipeptide carnosine, which is formed from one molecule of histidine and one of β-alanine.

$$H_2N - CH_2 - CH_2 - CO - \underset{\underset{H}{|}}{N} - \underset{\underset{CO_2H}{|}}{\overset{\overset{CH_2}{|}}{CH}}$$

Carnosine

A second possible strategy for dealing with metabolic acidosis is to try to increase the intracellular buffering capacity. The muscle cell has a number of buffer agents that can act within the physiological range, but some of these, such as phosphate, play other important roles in the cell. For an effective buffer, the dissociation constant (pKa) needs to lie within the physiological range: for muscle, this is about 6.5–7.1. Peptides and proteins can act as buffers, primarily because of the histidine residues that they contain: the pKa of the imidazole ring of histidine is about 6.1. If histidine is linked to a molecule of β-alanine to form the dipeptide carnosine, the pKa is increased to 6.83 (Figure 4.9). Carnosine is present in human skeletal muscle and the carnosine content of Type-II muscle fibres is higher than that of Type-I fibres. In human Type-IIx fibres, carnosine may account for up to 50% of physico-chemical buffering of H^+ produced by muscle in the pH range 7.1–6.5, but the overall contribution in mixed muscle is probably about 10–20%. High-intensity training increases the muscle carnosine content, and it can also be increased by ingestion of carnosine supplements. Unlike the α-amino acids that make up proteins, β-alanine availability is limited and probably limits the synthesis of carnosine. Short periods of supplementation of the diet with β-alanine can increase the muscle carnosine content. Studies have shown that daily supplementation with 6.4 g β-alanine in 8 divided doses will increase the muscle carnosine content by 60% after 4 weeks and by 80% after 10 weeks. There is some evidence that elevations of the muscle carnosine content increase cycle exercise performance at a constant rate with an expected work time of 2–3 min, increase incremental cycle exercise capacity, and extends isometric contraction time at 50% maximum isometric muscle force.

1. For exercise lasting more than a few seconds, ATP derived from the anaerobic metabolism of glucose (or glycogen) becomes available. Glycolysis is the name given to this pathway and the end-product of this series of reactions is pyruvate.

2. In the two major stages of glycolysis, glucose is first phosphorylated and cleaved to form two molecules of the three-carbon sugar glyceraldehyde 3-phosphate. The second stage involves the conversion of this into pyruvate, accompanied by the formation of ATP and reduction of NAD to NADH.

3. Glycolysis makes two molecules of ATP available for each molecule of glucose that passes through the pathway. If muscle glycogen is the starting substrate, three ATP molecules are generated for each glucose unit passing down the pathway.

4. For the reactions of glycolysis to proceed, pyruvate must be removed. In low-intensity exercise (when the rate at which energy is required can be met aerobically) pyruvate is converted to CO_2 and H_2O by oxidative metabolism in the mitochondria. In high-intensity exercise, the pyruvate is removed anaerobically by conversion to lactate. This simultaneously allows NAD to be regenerated from NADH.

5. Lactate accumulates in the muscle when the rate of anaerobic glycolysis exceeds the rate of flux through the pyruvate dehydrogenase reaction (which converts pyruvate to acetyl CoA). Lactate accumulation is accompanied by accumulation of hydrogen ions, which may interfere with muscle activation and contraction.

6. The total capacity of the glycolytic system for producing energy is large in comparison with the phosphagen system. A large part, but not all, of the muscle glycogen store can be used for anaerobic energy production during high-intensity exercise and will supply the major part of the energy requirement for maximum intensity efforts lasting from 20 s to 5 min.

7. Submaximal, high-intensity (non-steady-state) exercise can be sustained for durations approaching 5 min before fatigue is evident. Under these conditions carbohydrate oxidation can make a significant contribution to ATP production, but its relative importance is often underestimated.

8. Fatigue is an inevitable feature of high-intensity exercise and can be defined as the inability to maintain a given or expected power output or force. The onset of muscle fatigue has been associated with the disruption of energy supply, product inhibition, and factors preceding cross-bridge formation. It is likely to be a multifactorial process.

9. As with the sprinter, much of the training programme of the middle-distance athlete consists of repeated bouts of high-intensity effort with variable rest periods. There is therefore a need to ensure an adequate dietary carbohydrate intake during periods of intensive training to maintain the muscle glycogen stores.

10. There is good evidence that performance is impaired if events begin with a low muscle glycogen content and a few days of reduced training combined with a high carbohydrate diet should ensure an adequate muscle glycogen store before racing. Ingestion of a low carbohydrate diet is also associated with a metabolic acidosis and this has potentially negative effects on performance.

11. The ingestion of alkaline salts such as sodium bicarbonate in the hours before exercise has been shown to improve performance in events lasting 3–7 min. This is probably due to elevation of the pH of the extracellular space and its buffering capacity, which allows a faster removal of hydrogen ions from the intracellular space and therefore increases the amount of lactate (and thus energy from glycolysis) that can be produced by the muscle before a critically low pH is reached. Increasing the intracellular buffer capacity by ingestion of carnosine or β-alanine can improve performance in some models of short-term high-intensity exercise.

Selected further reading

Bangsbo J (1997). Physiology of muscle fatigue during intense exercise. In: *The clinical pharmacology of sport and exercise* (ed. Reilly T and Orme M). Amsterdam: Elsevier.

Bangsbo J et al. (1990). Anaerobic energy production and O_2 deficit-debt relationships during exhaustive exercise in humans. *Journal of Physiology* 422: 539–559.

Bender DA (2008) Energy Nutrition – the metabolism of carbohydrates and fats. In: *Introduction to nutrition and metabolism*. 4th edn. Boca Raton. CRC Press, pp 115–170

Bird SR et al. (1995). The effect of sodium bicarbonate ingestion on 1500-m racing time. *Journal of Sports Sciences* 13: 399–403.

Essen B (1978). Glycogen depletion of different fibre types in human skeletal muscle during intermittent and continuous exercise. *Acta Physiologica Scandinavica* 103: 446–455.

Gaitanos GC et al. (1993). Human muscle metabolism during intermittent maximal exercise. *Journal of Applied Physiology* 75: 712–719.

Greenhaff PL et al. (1994). The metabolic responses of human type I and II muscle fibres during maximal treadmill sprinting. *Journal of Physiology* 478: 149–155.

Hargreaves M (2005) Skeletal muscle carbohydrate metabolism during exercise. In: Hargreaves M and Spriet LL (ed.) *Exercise metabolism*, 2nd edn. Champaign, IL: Human Kinetics, pp. 29–44.

Juel C et al. (2004). Effect of high-intensity intermittent training on lactate and H^+ release from human skeletal muscle. *American Journal of Physiology (Endocrinology and Metabolism)* 286: E245–E251.

Maughan RJ and Greenhaff PL (1991). High-intensity exercise and acid-base balance: the influence of diet and induced metabolic alkalosis on performance. In: *Advances in nutrition and top sport* (ed. Brouns F). Basel: Karger, pp. 147–165.

Maughan RJ, Gleeson M, and Greenhaff PL (1997). *Biochemistry of exercise and training*. Oxford: University Press.

Maughan RJ et al. (1997). Diet composition and the performance of high-intensity exercise. *Journal of Sports Sciences* 15: 265–275.

Mohr M et al. (2007). Effect of two different intense training regimens on skeletal muscle ion transport proteins and fatigue development. *American Journal of Physiology* 292: R1594–R1602.

Osnes JB and Hermansen L (1972). Acid-base balance after maximal exercise of short duration. *Journal of Applied Physiology* 32: 59–63.

Wilkes D et al. (1983). Effect of acute induced metabolic alkalosis on 800-m racing time. *Medicine and Science in Sports and Exercise* 15: 277–280.

Williams MH (1997). *The ergogenics edge*. Champaign, IL: Human Kinetics.

The endurance athlete

5

Learning objectives

After studying this chapter, you should be able to …

1. describe the energy and oxygen cost of prolonged submaximal exercise;
2. understand the importance of aerobic power and the ability to use a high fraction of the maximum aerobic capacity in the performance of long-distance events;
3. appreciate the important contribution of fat oxidation to energy metabolism in prolonged exercise;
4. describe the process of fat breakdown (lipolysis);
5. describe the pathways of fat oxidation in the mitochondria;
6. describe the metabolism of pyruvate derived from glycolysis;

7. describe the tricarboxylic acid (TCA) cycle;

8. describe how the aerobic process of electron transport-oxidative phosphorylation regenerates ATP from ADP;

9. understand the mechanisms of regulation of fuel use during exercise, including the role of hormones and cytokines in controlling energy metabolism;

10. discuss the mechanisms of fatigue in prolonged submaximal exercise.

Introduction

> The ability to supply energy by aerobic metabolism is a key factor in success in middle- and long-distance events

Among runners, events from 1500 to 5000 m are usually classified as middle distance and events at 10 000 m or longer as distance races. Races at distances longer that the marathon (42.2 km, or 26.2 miles) are commonly termed ultradistance. Based on the physiological and biochemical demands of the events, the runners' distinction between middle distance and long distance is a realistic one, although their classification is based solely on the experiences of coaches and athletes. Table 5.1 shows the current world best performances for men and women at various distances. Many other sports also involve exercise on this same time scale. Most team games last for about 90 min and are therefore endurance events, although they also include short bursts of high-intensity activity, as discussed in Chapter 6. Some events last much longer, with the Ironman triathlon taking about 10–11 h for elite performers. In cycling, stage races typically last a few days, but the Tour de France involves more than 20 days of cycling and a total distance of about 4000 km.

At the longer distances, lasting about 30 min or more, anaerobic metabolism plays only a small role in energy supply. The ability to supply energy by aerobic metabolism is a key factor in success in middle- and long-distance events. In the

Table 5.1 Current (at 1 March 2010) world records (h:min:s) for men and women at various long-distances events. The remarkable thing is how little the pace (shown in parentheses in kilometres per hour) slows as the distance increases.

	Men	Women
Marathon (42.2 km)	2:03:59 (20.4 km/h)	2:15:25 (18.7 km/h)
100 km	6:13:33 (16.2 km/h)	6:33:11 (15.3 km/h)
100 miles (160 km)	11:30:51 (13.9 km/h)	14:29:44 (11.0 km/h)
24 h	303.5 km (12.6 km/h)	250.1 km (10.4 km/h)

longer events, successful performers are characterized by a high capacity to use fat as a fuel. This requires a highly developed cardiovascular system to supply oxygen to the working muscles and a high activity in the muscles of the enzymes involved in oxidative metabolism.

Energy supply

For the first minute of exercise anaerobic glycolysis is the main energy source, but aerobic metabolism dominates thereafter

In the last chapter, the total energy cost of a male athlete running at world record speed for 1500 m was estimated to be about 366 ml of oxygen per kg body weight, or 105 ml of oxygen per kg body mass per minute if a constant speed is assumed. The total energy cost for the whole race is therefore equivalent to that supplied by about 25.6 l of oxygen for a 70-kg runner (see Figure 4.1). The oxygen deficit incurred (i.e. the amount of energy supplied by anaerobic metabolism) in running at this speed is not more than about 5 l. In other words, the entire amount of energy that can be supplied by anaerobic pathways can be met by oxidative metabolism in less than 1 min by a well-trained endurance athlete. The rate of oxygen consumption increases only slowly after the gun, but the runner has reached racing speed almost immediately. For the first minute, anaerobic glycolysis is the main energy source, but aerobic metabolism dominates thereafter.

Exercise at an intensity corresponding to VO_{2max} cannot be sustained for longer than a few minutes

In events of longer duration, the overall contribution of anaerobic metabolism to energy production declines, and the requirement for a high anaerobic capacity is of correspondingly less importance. For most individuals, exercise at an intensity corresponding to VO_{2max} cannot be sustained for longer than a few minutes, although elite performers can run 5000-m races at an intensity at or close to VO_{2max}. The substantial oxygen deficit incurred in the first few minutes persists or even increases as the race progresses. Even at longer distances, anaerobic energy production occurs at intermediate points in a race if the pace is increased or during uphill running. Some anaerobic effort is also normally involved in the closing stages during a sprint finish. Most of the outstanding male performers at distance of up to 10 000 m are capable of running 1 mile in rather less than 4 min, and most of the best women can run 1500 m in close to that time.

The anaerobic power that can be produced by endurance runners in standardized laboratory tests is, as might be expected, fairly low compared with that of trained sprint athletes. Marathon runners demonstrate low isometric strength of the quadriceps muscles relative to sprinters, and also tend to have less muscle strength than healthy but untrained individuals. These observations are perhaps unsurprising in view of the smaller muscle mass of the muscles of distance runners and of the high proportion of Type-I fibres present in their muscles, but they may also reflect the training patterns of these athletes.

Aerobic power

A high capacity for oxidative metabolism is a prerequisite for success in endurance events

As the duration of running events increases, so does the proportion of the total energy demand that must be met by oxidative metabolism, and a high capacity for oxidative metabolism is a prerequisite for success in distance running. This also requires a highly developed cardiovascular system to ensure that the supply of oxygen to the muscles can keep pace with the rate at which it is being used. For each individual, there is a clearly defined limit to the maximum rate at which oxygen can be used. This is referred to as the maximum oxygen uptake (VO_{2max}) or maximum aerobic capacity. The limiting factor in most situations is the ability of the heart to pump oxygen-rich blood around the circulation, but there are some situations where the ability of the lungs to deliver enough oxygen to the blood may be the limiting factor. This applies especially at altitude, where the oxygen content of the air is reduced. The ability of the muscles to use oxygen normally exceeds the capacity of the circulatory system to supply it, but muscles of the successful endurance athlete have a high concentration of the enzymes necessary to supply energy by aerobic metabolism.

A high capacity for oxidative metabolism but does not in itself distinguish the elite performer

When comparisons are made within groups of runners of widely different levels of performance, a good relationship between endurance performance and VO_{2max} values is apparent. This is true for middle-distance events as well as for long-distance events. The relationship between finishing time in a marathon race and VO_{2max} measured within a few weeks of the race is shown in Figure 5.1 and clearly illustrates the greater aerobic capacity of the faster runners. Within groups of individuals of comparable levels of performance, however, there is no good relationship between VO_{2max} and performance, suggesting that a high capacity for oxidative metabolism is necessary for success in distance running, but does not in itself distinguish the elite performer. It is not unusual to find club athletes who can achieve VO_{2max} values of 70–75 ml/kg/min despite their relatively modest performances.

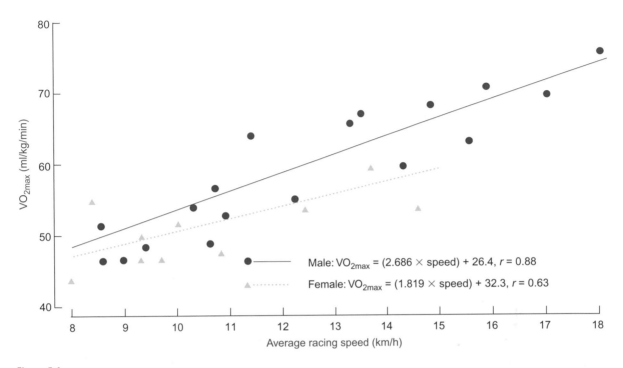

Male: $VO_{2max} = (2.686 \times speed) + 26.4$, $r = 0.88$

Female: $VO_{2max} = (1.819 \times speed) + 32.3$, $r = 0.63$

Figure 5.1

Relationship between finishing time in a marathon race and VO_{2max} measured within a few weeks of the race.

Fractional utilization of aerobic capacity

The best long-distance runners are characterized by an ability to run at a high fraction of their VO_{2max}

Running speed declines as the distance increases; based on current world records, it is apparent that the decrease in speed, and hence in energy demand, is gradual with increasing duration of effort (Figure 5.2). In middle-distance events, runners are working at intensities close to or above VO_{2max}. It has been estimated that trained athletes can sustain 100% of VO_{2max} for 10 min, 95% for 30 min, 85% for 60 min, and 80% for 120 min. In the longer events, where the demand is met almost entirely by aerobic metabolism, runners with a high VO_{2max} can meet the oxygen requirement by using a relatively low fraction of their maximum: runners who have a lower VO_{2max} have to work at a relatively higher intensity to run at the same speed. Part of the apparent lack of a close association between VO_{2max} and performance in long-distance races may be accounted for by differences between individuals in the fraction of VO_{2max} that can be sustained for the duration of a race. Although a good relationship between marathon running

Figure 5.2

Relationship between running speed based on current world records and the distance run in the race. It is apparent that the decrease in speed, and hence in energy demand, is gradual with increasing duration of effort.

performance and the fraction of VO_{2max} that can be sustained for the duration of the race is seen when runners of widely different levels of ability are compared (Figure 5.3), there is generally no such relationship seen when homogeneous groups are compared.

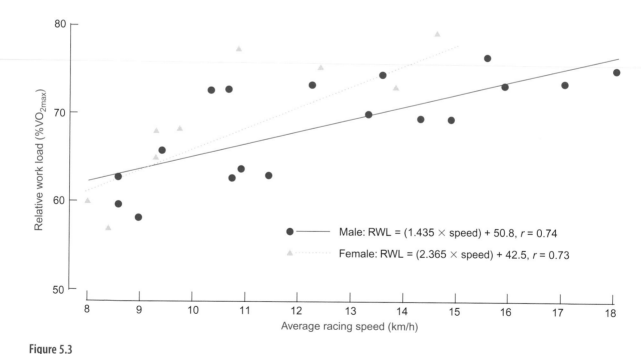

Male: RWL = (1.435 × speed) + 50.8, $r = 0.74$

Female: RWL = (2.365 × speed) + 42.5, $r = 0.73$

Figure 5.3

Relationship between marathon running performance and the fraction of VO_{2max} that can be sustained for the duration of the race for runners of widely different levels of ability.

The general trend does suggest that the fastest runners are characterized by an ability to run at a high fraction of VO_{2max} over any given distance. This is hardly surprising, as the fraction of VO_{2max} that can be maintained is more closely related to time than to distance, and the faster runners take less time to cover any given distance. For any individual, however, the fraction of VO_{2max} that can be sustained decreases with the distance. When the best performances of a group of highly trained marathon runners were compared, it was found that they used 94% of VO_{2max} when racing over 5 km (run time 15 min 49 s), 82% over 42.2 km (2 h 31 min), and 67% over 84.4 km (5 h 58 min). One of the major effects of endurance training is to increase the ability to utilize a large fraction of VO_{2max} for prolonged periods, and substantial improvements in racing performance can be achieved without any measurable change in VO_{2max}.

> The muscles of elite endurance athletes contain a high proportion of Type-I fibres

The ability of elite endurance athletes to sustain high rates of energy supply for very prolonged periods is in part a reflection of the high proportion of Type-I fibres present, reflecting a genetically determined predisposition to success in these events. The muscles of these individuals also contain rather few Type-IIX fibres, most of the fast-twitch fibres being high oxidative Type-IIa fibres. The activity of enzymes involved in oxidative metabolism is, however, generally high in both of the major fibre types in these individuals, reflecting an adaptation to the training programme. Such is the capacity for adaptation that the oxidative capacity of the Type-II fibres of the highly trained endurance athlete may exceed that of the Type-I fibres of the sedentary individual. There are still, however, distinct fibre types present in the trained muscle, and the oxidative capacity of Type-II fibres does not exceed that of Type-I fibres from the same individual. This is discussed in more detail later.

The major significance of the local adaptations within the muscle may be an increased capacity for use of fat as a fuel, leading to a slower rate of depletion of the limited muscle glycogen stores. At the same work intensity, the trained individual has a greater rate of fat oxidation, reflecting an increased delivery of blood-borne free fatty acids secondary to the increased capillary supply of the muscle as well as the enhanced capacity of the muscle to oxidize fat.

Energy metabolism

> The higher the exercise intensity the greater the energy demand and the greater the reliance on carbohydrate metabolism

The relative contributions of anaerobic and aerobic metabolism to energy production in events over different distances have been referred to earlier. The primary factor influencing the metabolic response to exercise is the intensity: the higher the intensity, the greater the energy demand and the greater the proportion of the total energy turnover that is met by carbohydrate metabolism. When the exercise intensity exceeds about 95% of VO_{2max}, the contribution of fat oxidation to energy metabolism is negligible. Using Snell's estimate of 5.7 l/min as the oxygen cost of running at 4-min mile pace for a 70-kg runner, and assuming that the entire energy demand (approximately 120 kJ/min) could be met by oxidation of muscle glycogen, it can be calculated that the rate of carbohydrate oxidation necessary to meet this rate of energy expenditure would be 7.5 g/min. Hence, a total of 30 g of carbohydrate would be used up by the end of the race. If, instead, the energy supply was met entirely by anaerobic glycolysis, the rate of carbohydrate degradation would be approximately 100 g/min. For a runner with a VO_{2max} of 70 ml/kg/min who can use 75% of that value in the first minute and 100% thereafter, and ignoring the contribution of phosphocreatine hydrolysis, the total carbohydrate degradation during the 1-mile race would be about 110 g. Of this amount, some 85 g would be converted to lactate, which, if distributed equally throughout 85% of the body water space, would reach a concentration of just over 26 mmol/l. A post-race blood lactate concentration of approximately 24 mmol/l has been recorded in an athlete who ran 1500 m at a slightly slower pace in 3 min 48 s, which is equivalent to about 4 min 6 s for 1 mile: a similar value was found in a runner who completed 5000 m in a time of 13 min 46 s.

Carbohydrate availability is widely recognized as a potential limitation to endurance exercise performance

These rough calculations show that the amount of muscle glycogen used is small relative to the whole body glycogen store (~500 g), and that substrate availability should not be limiting in events of this distance. There are, however, suggestions that the availability of muscle glycogen may limit the performance of events of this duration, perhaps because of depletion in specific muscle fibre pools. The implications of lactate accumulation for acid–base status and fatigue are discussed later.

At submaximal exercise intensities, the contribution of fat oxidation to energy production increases with time, but the contribution of fat to energy metabolism is likely to be insignificant at distances of less than 10 km run at race pace. Even at the marathon distance, where the energy demand can be met almost entirely by aerobic metabolism, the total amount of fat oxidized is small: if fat were the only fuel used, the total amount oxidized in a race would be no more than about 300 g. By contrast, if carbohydrate were the only fuel used, the total would be about 700 g, and this amount is considerably in excess of the amount that is normally stored in the liver and muscles.

> Blood glucose concentration is maintained during starvation and prolonged exercise by liver glycogen breakdown and gluconeogenesis

The liver plays a central role in the body's carbohydrate homeostasis. It acts as a reservoir of glycogen: the presence of the enzyme glucose 6-phosphatase allows free glucose to be released from the cell for the maintenance of the blood glucose concentration. The high capacity of liver tissue for gluconeogenesis also allows the liver to play a major role in providing glucose for the brain and other obligatory glucose users in times of starvation or carbohydrate deprivation. Even during short periods of starvation, as in the normal overnight fast, the liver glycogen content falls markedly and gluconeogenesis is accelerated. Gluconeogenesis is the synthesis of glucose from non-carbohydrate sources and becomes very important during starvation by making glucose available to those tissues (e.g. the brain and red blood cells) that cannot use any other fuel. Gluconeogenesis also allows recycling of the lactate produced by tissues such as red blood cells: the body's total lactate turnover per day may exceed 100 g, and lactate losses from the body in sweat and urine are extremely small. During and after exercise, if the blood lactate has been elevated, the liver removes a large part of this and uses it to resynthesize glucose. A variety of substrates can contribute to the process of gluconeogenesis. Lactate and pyruvate produced by glycolysis can be used, as well as the glycerol backbone of triglyceride molecules and the carbon skeletons of some amino acids, particularly alanine and glutamine, which are both released in increasing amounts from skeletal muscle during prolonged exercise.

In cycling at moderate intensities (about 70–75% of VO_{2max}) a progressive fall in the muscle glycogen takes place: the rate of depletion falls as the glycogen content itself falls, and at the point of fatigue, at least in cycling, levels close to zero are found (see Figure 5.4). Carbohydrate availability is widely recognized as a potential limitation to endurance performance, and carbohydrate ingestion during exercise

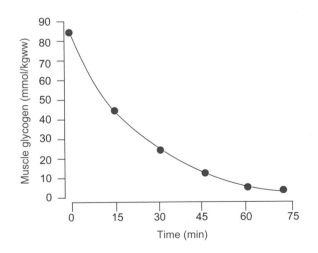

Figure 5.4

The glycogen content of exercising muscles falls progressively, and in cycling exercise at intensities that can be sustained for about 1–3 h, the point of fatigue occurs when a critically low level of glycogen is reached. Exercise at lower intensities is possible using fat as a fuel, but exercise cannot continue at higher intensities.

lasting more than about 60 min is effective in improving performance. Nutritional strategies to increase carbohydrate storage in muscle are therefore important for endurance athletes and are discussed at the end of this chapter.

Oxidation of carbohydrate

> The energy made available by anaerobic glycolysis allows the performance of high-intensity exercise that would otherwise not be possible

During glycolysis, NAD^+ is converted to NADH as described in Chapter 4. If the NADH formed by glycolysis is not reoxidized to NAD^+ at an equal rate, the rate of glycolysis—and thus of energy supply to the cell—will be reduced. At high rates of glycolysis, the availability of NAD^+ becomes limiting and is a potential cause of fatigue. As described in the previous chapter, reduction of pyruvate to lactate allows regeneration of NAD^+, and allows exercise to continue. This reaction has the advantage that it can proceed in the absence of oxygen, and is therefore not limited by the sluggishness of the cardiovascular system.

The negative effects of the acidosis resulting from lactate accumulation are often stressed, but it must be remembered that the energy made available by anaerobic glycolysis allows the performance of high-intensity exercise that would otherwise not be possible. Nonetheless, the amount of energy that can be made available in this way is small, and uses only a fraction of the chemical energy available in the glucose molecule.

As an alternative to conversion to lactate, pyruvate may undergo complete oxidation to CO_2 and water. This process occurs within the mitochondria of the cell (which accounts for the fact that red blood cells—which do not possess mitochondria—cannot meet any of their energy needs by oxidative metabolism). Pyruvate, which is produced in the cytoplasm, is transported across the mitochondrial membrane by a specific carrier protein. The first step to occur within the mitochondrion is an oxidative decarboxylation reaction catalysed by the enzyme pyruvate dehydrogenase (see Figure 5.5). This is a complex reaction involving the removal of one carbon atom (as carbon dioxide) from pyruvate and the conversion of another molecule of NAD^+ to NADH. In the process of converting the three-carbon pyruvate molecule to a two-carbon acetate group, a cofactor (coenzyme A) is added to form acetyl-CoA.

> Acetyl-CoA is oxidized to carbon dioxide and water in the tricarboxylic acid (TCA) or Krebs cycle

Acetyl-CoA is oxidized to carbon dioxide and water in the tricarboxylic acid (TCA) cycle: this series of reactions is also known as the Krebs cycle, after Hans Krebs who first described the reactions involved, or the citric acid cycle, as citrate

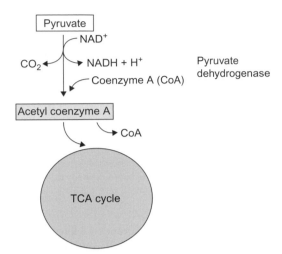

Figure 5.5

The first step in the metabolism of pyruvate following its entry into the mitochondrion is an oxidative decarboxylation reaction catalysed by the enzyme pyruvate dehydrogenase. This is a complex reaction, involving the removal of one carbon atom (as carbon dioxide) and the conversion of another molecule of NAD^+ to NADH. In the process of converting the three-carbon pyruvate molecule to a two-carbon acetate group a cofactor, coenzyme A, is added to form acetyl-CoA.

is one of the key intermediates in the process. The reactions involve combination of a molecule of acetyl-CoA with a four-carbon molecule, oxaloacetate, to form citrate, a six-carbon tricarboxylic acid. A series of reactions leads to the sequential loss of hydrogen ions and carbon dioxide, resulting in the regeneration of oxaloacetate. Acetyl-CoA is also a product of the beta-oxidation of fatty acids, and thus the final steps of oxidative degradation are common to both fat and carbohydrate. These reactions are described later in this chapter.

Fat as a fuel

Fat is a far more efficient storage form of energy than carbohydrate

The body has relatively large stores of fat, which is a highly concentrated energy storage form. Fat, protein, and carbohydrate (and alcohol) all release energy when oxidized, but as you can see in Table 5.2, the amount per gram of fuel varies greatly. In addition, the adipose cells in which the fat is stored contain little water, whereas storage of carbohydrate as glycogen in muscle means that 2–3 g of water are stored with each gram of carbohydrate. This makes fat a far more efficient storage form of energy than carbohydrate. Proteins can be broken down to provide energy, but all the proteins in the body serve functional or structural roles: there is no store of protein that can be broken down without some loss of functional capacity. Fat is therefore the only fuel that allows animals to lay down very large energy reserves that can sustain them through prolonged periods of starvation. Unlike carbohydrate, however, this energy cannot be made available at high rates through anaerobic metabolism but can only release energy fairly slowly in the process of being converted to carbon dioxide and water. Body fat stores vary greatly between individuals, but women generally have a larger fat store than men. In the average young man, fat accounts for about 15–20% of body mass, whereas young women

Table 5.2 Energy density of different macronutrients.

	kJ/g	kcal/g
Carbohydrate	16	3.75
Fat	37	9.0
Protein*	17	4.0
Alcohol	29	7.0

* The value for protein represents the available metabolizable energy. The total chemical potential energy available from protein is 22 kJ/g but in the body the nitrogenous component is converted to urea and is excreted. Although only a very small molecule, urea has an energy value of 5 kJ/g.

typically have about 25–33%. In elite endurance athletes this may be as low as 3–5% for men and 7–10% for women. This still amounts to a substantial energy reserve: the energy cost of running a marathon (about 12 MJ) could be met by the oxidation of only about 300 g of fat for a typical 70-kg runner. It would be a mistake to think of the fat cells as being nothing more than a storage depot. They protect vital organs and provide insulation from the cold. Other fats in the body play an essential role as components of membranes and nerve sheaths.

Fat metabolism

Fat stores

Triglyceride stores are found in adipose tissue and skeletal muscle

Almost all (about 90%) of the body's fat reserve is in the form of triglycerides, or triacylglycerols, stored in adipose cells, but a small and important part is stored in skeletal muscles. As the name implies, these molecules consist of three fatty acids (acyl molecules) bonded to a glycerol molecule, as shown in Figure 5.6. Each of the fatty acids consists of a backbone of carbon atoms, with hydrogen atoms bonded to each, and a carboxyl group (–COOH), which gives it its acidic properties, at one end. The number of carbons in the chain varies, but most contain an even number (between 14 and 22) of carbons; 16- and 18-carbon fatty acids are the most common.

Lipolysis

The breakdown of triglyceride into its fatty acid and glycerol components is called lipolysis

As mentioned earlier, acetyl-CoA is also a product of oxidative fatty acid degradation, and the sequence of reactions involving the TCA cycle and oxidative

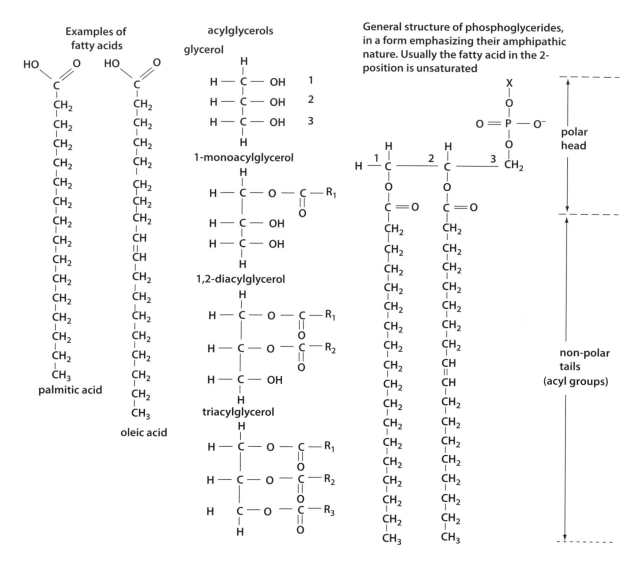

Figure 5.6

Triglyceride or triacylglycerol molecules consist of three fatty acids (acyl molecules) bonded to a glycerol molecule. Each of the fatty acids consists of a backbone of carbon atoms, with hydrogen atoms bonded to each, and a carboxyl group (–COOH), which gives it its acidic properties, at one end. The number of carbons in the chain varies, but most contain an even number (between 14 and 22) of carbons; 16- and 18-carbon fatty acids are the most common.

phosphorylation are therefore common to both fat and carbohydrate. In order to generate the two-carbon acetyl groups from fat, several metabolic steps have to occur. The first involves the breakdown of the storage form of fat, triglyceride, into its fatty acid and glycerol components. This process is called lipolysis and it begins with the hydrolytic removal of a fatty acid molecule from the glycerol backbone at either position 1 or position 3. This step, which is illustrated in

Figure 5.7

The process of lipolysis (the breakdown of triglyceride into its fatty acid and glycerol components) by the action of a hormone-sensitive lipase in adipose tissue. From each molecule of triglyceride, one molecule of glycerol and three molecules of free fatty acid (FFA) are produced.

Figure 5.7, is catalysed by a hormone-sensitive lipase. A specific lipase for the remaining diglyceride removes another fatty acid and another specific lipase removes the last fatty acid from the monoglyceride. Thus, from each molecule of triglyceride, one molecule of glycerol and three molecules of free fatty acid (FFA) are produced.

The rate of lipolysis and the rate of adipose tissue blood flow determine the rate of entry of FFA into the circulation

 Leaders in the field

Hans Krebs (1900–1981) studied medicine at the Universities of Göttingen, Freiburg-im-Breisgau, and Berlin and then spent one year studying chemistry at Berlin. In 1926 he was appointed Assistant to Professor Otto Warburg at the Kaiser Wilhelm Institute for Biology at Berlin-Dahlem, where he remained until 1930. His appointment was terminated by the National Socialist Government in 1933, and Krebs was invited by Sir Frederick Gowland Hopkins, to the School of Biochemistry, Cambridge. In 1935, he was appointed Lecturer in Pharmacology at the University of Sheffield, and in 1938 Lecturer-in-Charge of the Department of Biochemistry then newly founded there. In 1945 this appointment was raised to that of Professor, and of Director of a Medical Research Council's research unit in Human Metabolism. In 1954 he was appointed Whitley Professor of Biochemistry in the University of Oxford and the Medical Research Council's Unit for Research in Cell Metabolism was transferred to Oxford. Professor Krebs' research was mainly concerned with various aspects of intermediary metabolism, and he worked on the synthesis of urea in the liver. He is perhaps best known for another metabolic cycle, widely known as the Krebs cycle (otherwise referred to as the tricarboxylic acid cycle), where he identified the intermediary stages in the oxidation of fat and carbohydrate occurring in the mitochondria.

Both glycerol and FFA pass out of the adipose cells and into the circulation. Both the rate of lipolysis and the rate of adipose tissue blood flow influence the rate of entry of FFA and glycerol into the circulation. It is worth noting here that during prolonged exercise at about 50% VO_{2max} adipose tissue blood flow is increased. However, during intense exercise, sympathetic vasoconstriction results in a fall in adipose tissue blood flow, resulting in accumulation of FFA within adipose tissue, and effectively limiting the entry of FFA (and glycerol) into the circulation. Another factor that limits fat mobilization during high-intensity exercise is the accumulation of lactate in the blood. Lactate promotes the re-esterification of FFA produced during lipolysis (see Figure 5.7) and therefore limits the entry of FFA into the bloodstream.

Transport of fatty acids in blood and uptake by muscle

Most of the free fatty acids in plasma are transported loosely bound to albumin

Glycerol in plasma can be taken up by the liver and phosphorylated to glycerol 3-phosphate, which can be used to form triglyceride, as described earlier, or alternatively can be oxidized to dihydroxyacetone phosphate, which can enter either the glycolytic or gluconeogenic pathway. Free fatty acids are very poorly soluble in water, and most of the fatty acids in plasma are transported loosely bound to the plasma protein, albumin. The usual resting plasma concentration of FFA is 0.2–0.4 mmol/l. However, during (or shortly after) prolonged exercise the plasma FFA concentration may rise to about 2.0 mmol/l.

Uptake of FFA by muscle is directly related to the plasma FFA concentration and hence the lipolytic mobilization of lipid stores is an important step in ensuring an adequate nutrient supply for prolonged muscular work. The normal plasma albumin concentration is about 45 g/l (approximately 0.7 mmol/l). Each albumin molecule contains three high-affinity binding sites for FFA (and seven other low-affinity binding sites). When the three high-affinity binding sites are full (at an FFA concentration of 2.0 mmol/l or more), the concentration of FFA not bound to albumin increases markedly, forming fatty acid micelles that are potentially damaging to tissues because of their detergent-like properties.

FFA cross the sarcolemmal membrane by a carrier-mediated transport mechanism

Fatty acid transport across the sarcolemmal membrane into the muscle fibre occurs by facilitated diffusion using a fatty acid transporter (FAT/CD36) protein that becomes saturated at high plasma unbound fatty acid concentration (associated with a total fatty acid concentration equivalent to about 1.5 mmol/l). FFA uptake into muscle only occurs if the intracellular FFA concentration is less than that in true aqueous solution in the extracellular fluid (namely <10 μmol/l). The low intracellular FFA concentration is probably maintained by the presence of a fatty acid binding protein (FABP) inside the cell, perhaps similar to those found in the cells of the small intestine and liver. It also follows that the rate of uptake of FFA into muscle fibres will be proportional to the difference in their concentrations inside and outside the cell, up to a limit when the plasma membrane transport mechanism becomes saturated. After the fatty acids enter the muscle cell, they are converted to a CoA derivative by the action of ATP-linked fatty acyl-CoA synthetase (also known as thiokinase), in preparation for β-oxidation, the major pathway for fatty acid breakdown. Hence, the priming (activation) of each fatty acid molecule requires the utilization of one molecule of ATP:

$$RCOOH + ATP + CoA\text{-}SH \rightarrow R\text{-}C(=O)\text{-}S\text{-}CoA + AMP + PP_i.$$

Transport of fatty acids from the sarcoplasm into the mitochondria

Carnitine is needed to transport fatty acids across the mitochondrial membrane

Fatty acids are only broken down and oxidized in the mitochondria. However, fatty acyl-CoA molecules cannot simply diffuse across the mitochondrial outer and inner membranes. As you can see in Figure 5.8, fatty acyl-CoA molecules in the muscle sarcoplasm are transported into the mitochondria via formation of an ester of the fatty acid with carnitine. The latter is synthesized in the liver and is normally abundant in tissues able to oxidize fatty acids. Carnitine concentrations of about 1.0 mmol/l are found in human skeletal muscle. L-Carnitine is derived from the diet (red meats and dairy products) and from endogenous production in the body. Even when a diet is carnitine deficient, healthy humans are still able to produce enough carnitine from the amino acids methionine and lysine to maintain functional body stores. For this reason, carnitine is not regarded as a vitamin, but as a vitamin-like substance. The synthesis of L-carnitine occurs in liver and kidney, which together contain less than 2% of the whole body carnitine store (about 27 g). About 98% of the carnitine in the human body is present in skeletal muscle and heart. Skeletal muscle and the heart are dependent upon transport of L-carnitine through the circulation. Only about 0.5% of whole body carnitine is found circulating in the blood (the plasma carnitine concentration is about 0.05 mmol/l). Muscle takes up carnitine against a very large concentration gradient by an active transport process. The enzyme regulating the transport of FFA via carnitine is called carnitine acyltransferase (CAT). As you can see in Figure 5.8, two forms of the enzyme exist in muscle: one (CAT1) is located on the outer surface of the membrane (to generate acyl-carnitine) and the other (CAT2) is located on the inner surface of the inner mitochondrial membrane and regenerates the acyl-CoA and free carnitine. This transport process may be the main rate-limiting step in the utilization of fatty acids for energy production in muscle.

Figure 5.8

Fatty acyl-CoA molecules in the muscle sarcoplasm are transported into the mitochondria via formation of an ester of the fatty acid with carnitine. The enzyme regulating the transport of free fatty acid via carnitine is called carnitine acyltransferase and two forms of the enzyme exist in muscle: one is located on the outer surface of the membrane (to generate acyl-carnitine) and the other is located on the inner surface of the inner mitochondrial membrane and regenerates the acyl-CoA and free carnitine. This transport process may be the main rate-limiting step in the utilization of fatty acids for energy production in muscle.

Energy cannot be derived from fat via anaerobic pathways

At high exercise intensities (above about 60% VO_{2max}) the rate of fatty oxidation cannot provide sufficient ATP for muscle contraction, and so ATP is derived increasingly from carbohydrate oxidation and anaerobic glycolysis. Energy cannot be derived from fat via anaerobic pathways. Once released into the mitochondrial matrix, the fatty acyl-CoA is able to enter the β-oxidation pathway. Carnitine acyltransferase is inhibited by malonyl-CoA, a precursor for fatty acid synthesis. Hence, when the ATP supply is sufficient, surplus acetyl-CoA is diverted away from the TCA cycle to malonyl-CoA, hence reducing catabolism of fatty acids and promoting their formation and subsequent triglyceride synthesis.

β-Oxidation of fatty acids

Fatty acids are broken down to acetyl-CoA in the mitochondria in a process called β-oxidation

The process of β-oxidation occurs in the mitochondria and is the sequential removal of two-carbon units in the form of acetyl-CoA from the fatty acid chain of the fatty acyl-CoA molecule. This series of reactions is catalysed by a multienzyme complex that releases a molecule of acetyl-CoA and a fatty acyl-CoA, which is now a two-carbon unit shorter. This fatty acyl-CoA can now repeat the cycle, while the acetyl-CoA formed can enter the TCA cycle. At each passage through the cycle, the fatty acid chain loses a two-carbon fragment as acetyl-CoA and two pairs of hydrogen atoms to specific acceptors. The 16-carbon palmitate molecule thus undergoes a total of seven such cycles, to yield in total eight molecules of acetyl-CoA and 14 pairs of hydrogen atoms. The palmitate only needs to be primed or activated with CoA once, because at the end of each cycle the shortened fatty acid appears as its CoA ester. The most common fatty acids oxidized contain 16 (e.g. palmitate) or 18 (e.g. oleate) carbons in the acyl chain. The 14 pairs of hydrogen atoms removed during β-oxidation of palmitate enter the mitochondrial respiratory chain (described later in this chapter), seven pairs in the form of the reduced flavin coenzyme of fatty acyl-CoA dehydrogenase and seven pairs in the form of NADH. The passage of electrons from $FADH_2$ to oxygen and from NADH to oxygen leads to the expected number of oxidative phosphorylations of ADP (namely two ATP from each $FADH_2$ and three ATP from each NADH). Hence, a total of five molecules of ATP are formed per molecule of acetyl-CoA cleaved:

$$\text{Palmitoyl-CoA} + 7\,\text{CoA} + 7\,O_2 + 35\,\text{ADP} + 35\,P_i \rightarrow 8\,\text{acetyl-CoA} + 35\,\text{ATP} + 42\,H_2O.$$

Acetyl-CoA formed from the breakdown of fatty acids (or carbohydrate) is oxidized in the TCA cycle

The eight molecules of acetyl-CoA can enter the TCA cycle and the following equation represents the balance sheet for their oxidation and the coupled phosphorylations:

$$8\,\text{acetyl-CoA} + 16\,O_2 + 96\,\text{ADP} + 96\,P_i \rightarrow 8\,\text{CoA} + 96\,\text{ATP} + 104\,H_2O + 16\,CO_2.$$

Combining the two equations above gives the overall equation:

$$\text{Palmitoyl-CoA} + 23\,O_2 + 131\,\text{ADP} + 131\,P_i \rightarrow \text{CoA} + 16\,CO_2 + 146\,H_2O + 131\,\text{ATP}.$$

As one molecule of ATP was required to activate the free fatty acid to begin with, the net yield for the complete oxidation of one molecule of palmitic acid is 130 molecules of ATP.

The tricarboxylic acid cycle

The most important function of the TCA cycle is to generate hydrogen atoms for their subsequent passage to the electron-transport chain by means of NADH and $FADH_2$

The common product of both carbohydrate and fatty acid degradation, acetyl-CoA, is oxidized to CO_2 in the TCA cycle, which takes place within the inner

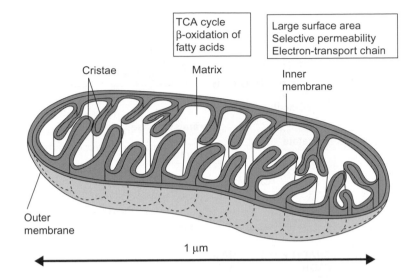

TCA cycle
β-oxidation of
fatty acids

Large surface area
Selective permeability
Electron-transport chain

Cristae   Matrix   Inner membrane

Outer membrane

1 μm

Figure 5.9

Diagram showing a cross-section through a mitochondrion. The folding of the inner membrane into finger-like projections called cristae gives a large surface area. The components of the electron-transport chain are embedded within the inner membrane. The enzymes of the TCA cycle and β-oxidation are found within the inner matrix.

matrix of the mitochondria (Figure 5.9). The reactions involve combination of acetyl-CoA with oxaloacetate to form citrate, a six-carbon tricarboxylic acid. A series of reactions leads to the sequential loss of hydrogen atoms and CO_2, resulting in the regeneration of oxaloacetate:

$$\text{acetyl-CoA} + \text{ADP} + P_i + 3\,\text{NAD}^+ + \text{FAD} + 3\,\text{H}_2\text{O} \rightarrow 2\,\text{CO}_2 + \text{CoA} + \text{ATP} + 3\,\text{NADH} + 3\text{H}^+ + \text{FADH}_2.$$

The hydrogen atoms are carried by the reduced coenzymes NADH and flavin adenine dinucleotide ($FADH_2$). These act as carriers and donate pairs of electrons to the electron-transport chain, allowing oxidative phosphorylation with the subsequent regeneration of ATP from ADP.

A summary of the reactions involved in the TCA cycle is shown in Figure 5.10. Note that molecular oxygen (O_2) does not participate directly in the reactions of the TCA cycle. In essence, the most important function of the TCA cycle is to generate hydrogen atoms for their subsequent passage to the electron-transport chain by means of NADH and $FADH_2$.

The terminal respiratory system

For each molecule of NADH that enters the electron-transport chain, three molecules of ATP are generated, and for each molecule of $FADH_2$, two molecules of ATP are formed

The aerobic process of electron transport-oxidative phosphorylation regenerates ATP from ADP, thus conserving some of the chemical potential energy contained within the original substrates in the form of high-energy phosphates.

Figure 5.10

(a) A simplified summary of the reactions involved in the TCA cycle and (b) the same pathway shown in more detail. Note that molecular oxygen (O_2) does not participate directly in the reactions of the TCA cycle. In essence, the most important function of the TCA cycle is to generate hydrogen atoms for their subsequent passage to the electron-transport chain by means of the reduced coenzymes NADH and $FADH_2$.

(a)

(b)

ENZYMES: 1. Citrate synthase 2. Aconitase 3. Aconitase 4. Isocitrate dehydrogenase 5. α-Ketoglutarate dehydrogenase 6. Succinyl CoA synthetase 7. Succinate dehydrogenase 8. Fumarase 9. Malate dehydrogenase

As long as there is an adequate supply of O_2, and substrate is available, NAD^+ and FAD are continuously regenerated and TCA metabolism proceeds. This system cannot function without the use of oxygen. For each molecule of NADH that enters the electron-transport chain, three molecules of ATP are generated, and for each molecule of $FADH_2$, two molecules of ATP are formed. Thus, for each molecule of acetyl-CoA undergoing complete oxidation in the TCA cycle, a total of 12 ATP molecules are formed.

The transfer of electrons through the electron-transport chain located on the inner mitochondrial membrane causes hydrogen ions or protons (H^+) from the inner mitochondrial matrix to be pumped across the inner mitochondrial membrane into the space between the inner and outer mitochondrial membranes. The high concentration of positively charged hydrogen ions in this outer chamber cause the H^+ ions to flow back into the mitochondrial matrix through an ATP synthase protein complex embedded in the inner mitochondrial membrane. The flow of H^+ ions (protons) through this complex constitutes a proton-motive force that is used to drive ATP synthesis.

Amino acids as a fuel source for exercise

Some amino acids can be oxidized by muscle

In addition to carbohydrate and fat, skeletal muscle can also utilize some amino acids as a source of fuel for exercise. Before amino acids can be oxidized, the amino group must be removed. Removal of the amino group can be achieved for some amino acids by transferring it to another molecule called a keto-acid, which results in the formation of a different amino acid. This process is called transamination and is catalysed by enzymes called aminotransferases. A good example is the transfer of the amino group from the amino acid leucine to the keto-acid α-ketoglutarate forming α-ketoisocaproate (which can be further metabolized to form acetyl-CoA) and glutamate, respectively. Each amino acid has its own, unique corresponding keto-acid. Alternatively, the amino group can be removed from the amino acid to form free ammonia (NH_3), in a process called oxidative deamination. One example is the breakdown of asparagine to form aspartate and NH_3. Because free ammonia is a toxic substance, it is used either to form glutamine from glutamate within the muscle or is transported as glutamine or alanine to the liver, where it is converted to urea and is eventually excreted by the kidneys. Both ammonia and urea can be excreted in urine and sweat.

After the removal of the amino group from an amino acid, the remaining carbon skeleton (the keto-acid) is eventually oxidized to CO_2 and H_2O in the TCA cycle. The carbon skeleton of amino acids can enter the TCA cycle in several ways. Some can be converted to acetyl-CoA and enter the TCA cycle just like acetyl-CoA from carbohydrate or fat. They can also enter the TCA cycle as α-ketoglutarate or oxaloacetate as metabolites of glutamate and aspartate, respectively.

Some amino acids can serve as glucogenic or lipogenic (ketogenic) precursors. The amino acids that can be converted into α-ketoglutarate, oxaloacetate, or pyruvate can also be used for the synthesis of glucose in the liver (gluconeogenesis). Amino acids or keto-acids that are eventually broken down to acetyl-CoA can also be used in the synthesis of FFAs. Acetyl-CoA units can be used in an elongation process to form 'longer' FFAs by adding on to the hydrocarbon chain of the 16-carbon FFA, palmitate. The following are examples of some common aminotransferase reactions:

L-glutamate + oxaloacetate → α-ketoglutarate + L-aspartate

L-alanine + β-ketoglutarate → pyruvate + L-glutamate.

Several amino acids undergo reversible transamination. These amino acids include alanine, aspartate, glutamate, and the BCAAs: leucine, isoleucine, and valine. The BCAAs are the only essential amino acids that can undergo transamination. Transamination is usually rapid, and the main limiting factor is the fact that these processes sometimes take place in different tissues. Amino acids, thus, have to be transported via the circulation. Glutamate serves a central role in these transamination reactions because several amino acids (including leucine, valine, isoleucine, alanine, aspartate, and glutamate) can undergo transamination with glutamate.

Most of the nitrogen from amino acid degradation is transferred to α-ketoglutarate to form glutamate and subsequently glutamine; these two amino acids (glutamate and glutamine) are the most abundant free amino acids in muscle. Although all carbon skeletons of amino acids can be used for oxidation, only 6 of the available 20 amino acids in protein are oxidized in significant amounts by muscle: asparagine, aspartate, glutamate, isoleucine, leucine, and valine.

Amino acid oxidation has been estimated to contribute up to only about 15% to energy expenditure in resting conditions. During exercise, this relative contribution likely decreases because of an increasing importance of carbohydrate and fat as fuels. During very prolonged exercise, when carbohydrate availability becomes limited, amino acid oxidation may increase somewhat but the contribution of protein to energy expenditure decreases to a maximum of about 10% of total energy expenditure.

Integration and regulation of fuel use

Thirty-eight molecules of ATP are resynthesized for each molecule of glucose oxidized

In terms of the energy conservation of aerobic glucose metabolism, the overall reaction starting with glucose as the fuel can be summarized as follows:

$$\text{Glucose} + 6\,O_2 + 38\,\text{ADP} + 38\,P_i \rightarrow 6\,CO_2 + 6\,H_2O + 38\,\text{ATP}.$$

The total ATP synthesis of 38 moles per mole of glucose oxidized are accounted for primarily by oxidation of reduced coenzymes in the terminal respiratory system as follows:

ATP synthesized	Source
2	Glycolysis
6	NADH by glycolysis
24	NADH
4	$FADH_2$
2	GTP
Total = 38	

There is some evidence that the net yield of ATP from oxidative metabolism is not quite constant, but depends on which shuttle system is used to transport the NADH produced in the cytosol into the mitochondrion. Glucose gives 38 molecules of ATP if the malate/aspartate shuttle is used (39 ATP with glycogen). However, if the glycerol phosphate shuttle is used then two NADH molecules in the cytosol appear as two $FADH_2$ molecules inside the mitochondrion. The final yield of ATP is therefore four from the glycerol phosphate shuttle as opposed to six from the malate/aspartate shuttle. In skeletal muscle, the glycerol phosphate shuttle appears to predominate, while, the malate/aspartate shuttle predominates in the heart.

For fats, the equation will vary depending on the fatty acid being oxidized, but complete oxidation of palmitate is as follows:

$$Palmitate + 23\,O_2 + 130\,ADP + 130\,P_i \rightarrow 16\,CO_2 + 146\,H_2O + 130\,ATP.$$

Factors affecting key enzymes

> The pyruvate dehydrogenase complex catalyses the rate-limiting step in carbohydrate oxidation

Conversion of pyruvate to acetyl-CoA by the pyruvate dehydrogenase complex is the rate-limiting step in carbohydrate oxidation and is stimulated by an increased intracellular concentration of calcium, and decreased ratios of ATP/ADP, acetyl-CoA/free CoA and NADH/NAD$^+$, and thus offers another site of regulation of the relative rates of fat and carbohydrate catabolism. If the rate of formation of acetyl-CoA from the β-oxidation of fatty acids is high, as after 1–2 h of submaximal exercise, then this could reduce the amount of acetyl-CoA derived from pyruvate, cause accumulation of phosphoenol pyruvate and inhibition of PFK, thus slowing the rate of glycolysis and glycogenolysis. This forms the basis of the 'glucose–fatty acid cycle', proposed by Randle and colleagues, that has for many years been accepted to be the key regulatory mechanism in the control of carbohydrate and fat utilization by skeletal muscle. However, recent

Figure 5.11

Relative contributions of fat and carbohydrate (muscle glycogen and blood glucose) to energy supply at increasing levels of exercise. At low intensities of exercise, fat is the predominant fuel, but at high-intensity exercise most of the energy demand is met by carbohydrate metabolism.

work has challenged this hypothesis and it seems likely that the regulation of the integration of fat and carbohydrate catabolism in exercising skeletal muscle must reside elsewhere (for example at the level of glucose uptake into muscle, glycogen breakdown by phophorylase, or the entry of fatty acids into the mitochondria).

A key regulatory point in the TCA cycle is the reaction catalysed by citrate synthase. The activity of this enzyme is inhibited by ATP, NADH, succinyl-CoA, and fatty acyl-CoA; the activity of the enzyme is also affected by citrate availability. Hence, when cellular energy levels are high, flux through the TCA cycle is relatively low, but can be greatly increased when ATP and NADH utilization is increased, as during exercise.

It is important to recognize that integration of substrate use is a dynamic process. It changes with substrate availability, with exercise intensity, and with time. Increasing or decreasing either fat or carbohydrate availability will result in compensatory increases in the contribution of the other energy source to energy supply. Figure 5.11 shows the pattern of substrate use with increasing exercise intensity. At low exercise intensities, fat oxidation supplies most of the energy demand, but as the exercise intensity increases, so the contribution of carbohydrate, and especially of muscle glycogen, to energy provision increases. When the exercise intensity is close to 100% VO_{2max}, almost all of the energy demand is met by carbohydrate metabolism. In exercise at a constant power output, it takes time to mobilize fatty acids from the adipose tissue depots, and muscle glycogen is the main fuel used in the early stages when the glycogen content of the muscle is relatively high (Figure 5.4). Over time, the glycogen content falls and the rate of glycogen breakdown also falls as fatty acid oxidation increases to meet the energy demand.

Leaders in the field

John Holloszy has made enormous contributions in many areas of exercise biochemistry in a career that stretches back almost 50 years. His research interests have focused on the interactions of exercise, diet, and metabolic regulation. Several of his papers on the metabolic adaptations in skeletal muscle in response to training and disuse have become classics in the field, and many of his graduate students and post-docs have gone on to make major contributions to our understanding of exercise biochemistry. John has also been a pioneer in understanding the impact of exercise training on the quality of life of the elderly. In recent years, he has become one of the leading figures in gerontology research, and his studies include investigations into whether exercise can reverse some of the loss of functional capacity that has traditionally been considered an inevitable part of ageing. As early as the 1960s he engaged in studies looking at the metabolic and physical consequences of restriction of food intake and a recent study from his laboratory has shown that severe energy restriction reduces the risk of heart attack, stroke, and diabetes, suggesting other ways of combating some of the diseases of modern life.

Hormonal responses to exercise

Hormonal changes during exercise are important in mobilizing fuel, regulating cardiovascular changes, and maintaining thermoregulation and fluid balance. During exercise, the activity in the sympathoadrenal neuroendocrine system, and accordingly the plasma concentrations the catecholamines, noradrenaline, adrenaline, and dopamine, increase in relation to the oxygen uptake, although at any given level of oxygen uptake, catecholamine levels are generally higher during work with smaller muscle groups than with larger ones. The increase in sympathoadrenal activity is of major importance for the cardiovascular adaptation and thermoregulation in exercise. Both the heart rate and force of contraction of the ventricles are enhanced, the tone of arterioles in the gut, liver, kidneys, and non-contracting muscles and of veins is increased, and the spleen is made to contract. In this way, cardiac output is enhanced and blood volume and flow redistributed in favour of working muscles and the skin. Stimulation of sweat glands increases evaporative heat loss and sweating results in loss of water and electrolytes from the body. However, sympathoadrenal activity also has effects on water and electrolyte excretion via the kidneys, reducing such losses in exercise. By decreasing renal blood flow and changing its distribution, and by direct effects on renal tubular reabsorption of sodium and water, sympathoadrenal activity diminishes urinary sodium and water losses.

Furthermore, during exercise sympathoadrenal activity, mainly via β-adrenergic mechanisms, releases renin from the kidneys that converts angiotensinogen in the plasma to the biologically active angiotensin II that in turn promotes the secretion of aldosterone and antidiuretic hormone (ADH) and, accordingly, renal sodium and water excretion is reduced. These effects are designed to maintain blood volume and pressure. The most important effect of sympathoadrenal activity on hormone secretion in exercise is an α-adrenergic inhibition of insulin release that is exerted mostly by sympathetic nerves rather than elevated circulating levels of adrenaline. An exercise-induced increase in plasma glucagon levels is only seen when exercise is prolonged and is caused directly by a fall in the blood glucose concentration. Finally, adrenergic activity exerts an inotropic effect in skeletal muscle, as it does in the heart.

The plasma concentrations of the anterior pituitary hormones, growth hormone (GH), adrenocorticotrophic hormone (ACTH), prolactin (PRL), and thyroid-stimulating hormone (TSH) may increase during exercise. The ACTH increase results in an increase in cortisol secretion. Despite some increase in TSH increases in thyroid hormones (T3 and T4) are not commonly observed. Furthermore, the levels of mitochondrial enzymes dependent on these hormones, and basal metabolic rate are essentially not changed by exercise.

In man, the levels of follicle-stimulating hormone (FSH) and luteinizing hormone (LH) are unchanged during exercise. Nevertheless, increases in the concentrations of gonad hormones may be seen. These are probably due to haemoconcentration and decreases in clearance (because of decreased blood flow to liver and kidneys) and some increase in plasma testosterone levels are commonly observed during high-intensity exercise. In men, testosterone levels may decrease below levels at rest in the post-exercise period. Prolonged periods of intensive training may decrease basal testosterone and multifactorial stress that includes daily exercise has been shown to depress testicular function and to cause a number of other complex endocrine changes.

Motor centres in the brain and working muscles stimulate release of hypothalamic releasing factors and pituitary hormones

At the onset of exercise, impulses from motor centres in the brain as well as from working muscles elicit a work-load-dependent increase in sympathoadrenal activity and in release of some pituitary hormones (GH, ACTH, PRL, possibly TSH). These changes control the changes in secretion of subordinate endocrine cells: increased sympathetic nervous system activity depresses insulin secretion, stimulates release of adrenaline and noradrenaline from the adrenal medulla, and stimulates the renin-angiotensin-ADH system, and possibly the secretion of PTH and gastrin as well. ACTH stimulates cortisol secretion from the adrenal cortex. The mechanisms of altered hormonal secretion during exercise are summarized in Figure 5.12.

Figure 5.12

Mechanisms of altered hormone secretion during exercise.

The hormonal response to exercise depends on the relative rather than on the absolute work

The magnitude of the hormonal response to exercise is influenced by a number of factors. The hormonal response to exercise depends on the relative rather than on the absolute work intensity (i.e. the response at a given oxygen uptake in ml/min being lower if the individual has a maximal oxygen uptake that is high rather than if it is low. Endurance training in previously sedentary individuals has been shown to attenuate the hormonal response to exercise at the same absolute or relative exercise intensity. Furthermore, experiments with fasted healthy subjects, diabetics, and obese subjects have indicated that the concentrations of other hormones are higher during exercise when insulin availability is lower than normal in the time preceding exercise. Glucose administered by infusion or oral ingestion markedly blunts many hormonal responses to exercise including the magnitude of increases in circulating levels of catecholamines, GH, ACTH, and cortisol. The hormonal response to exercise is also influenced by environment (hormonal responses are generally greater for exercise at altitude than at sea level and at high temperatures compared with temperate conditions) and finally hormonal responses may be influenced by disease (e.g. blunted sympathoadrenal response to exercise is observed in asthmatic subjects). If exercise is continued, the hormonal changes may be gradually intensified by impulses from receptors sensing changes in blood glucose concentrations, body temperature, intravascular volume, and O_2 tension. For competitive athletes, hormonal response to exercise is usually greater during competition because of the added impact of psychological stress.

The hormonal response to exercise is essentially non-discriminating, promoting mobilization of both glycogen and triglyceride from extra- as well as intramuscular stores. Breakdown of the latter is also enhanced directly by contractions and certainly during very high-intensity exercise (e.g. sprinting) this is quantitatively far more important than any hormonal influence on fuel mobilization. The quantity of fuel mobilized from the different stores depends not only on the hormonal response and contractile activity, but also on the size of the fuel depots as well as on the state of hormone receptors and capacity of involved enzymes in these depots. Mobilization of extramuscular fuel is regulated in a 'feedforward' fashion rather than by 'feedback' mechanisms coupled to the minimal needs of the exercising muscles. In accordance with this view, it has been shown in running rats that liver glycogen is broken down and glucose forced on the muscles at a time at which they are rich in endogenous fuel. Furthermore, studied at the same exercise intensity, oxygen uptake, and muscular glycogen concentration, liver glucose production was higher in experiments in which liver glycogen stores had been increased by fructose than in experiments in which hepatic glycogen had been depleted before exercise. Mobilization of extra- and intramuscular fuel is probably neither accurately nor optimally matching the energy demands of the contractile machinery. Exact regulation would hardly be compatible with the fact that during exercise changes in the plasma levels of hormones influencing metabolism are not geared exclusively to metabolic needs, but are also adjusted to serve control of circulation, volume, and osmolality of body fluids, and temperature. Rather, one would expect that, to some extent, mobilized energy sources after a shorter or longer travel would be 'buffered' by re-storage. In fact, there is evidence that mobilization of energy sources in response to exercise may exceed energy expenditure, some of the mobilized fuel being accumulated extracellularly or stored in intra- or extramuscular depots.

Regulation of energy metabolism by hormones

Many hormones influence energy metabolism in the body. During exercise the interaction between insulin, glucagon, and the catecholamines (adrenaline and noradrenaline) is mostly responsible for fuel substrate availability and utilization; cortisol and growth hormone also have some significant effects.

Insulin inhibits lipolysis and promotes glucose uptake

Insulin is secreted by the β cells of the islets of Langerhans in the pancreas. Its basic biological effects are to inhibit lipolysis and increase the uptake of glucose from the blood by the tissues, especially skeletal muscle, liver, and adipose tissue; the cellular uptake of amino acids is also stimulated by insulin. These effects reduce the plasma glucose concentration, inhibit the release of glucose from the liver, promote the synthesis of glycogen (in liver and muscle), promote synthesis of lipid and inhibit FFA release (in adipose tissue), increase muscle amino acid uptake, and enhance protein synthesis. The primary stimulus for increased insulin secretion is a rise in the blood glucose concentration (e.g. following a meal). Exercise usually results in a fall in insulin secretion.

Glucagon has opposite effects to those of insulin

Glucagon is secreted by the α cells of the pancreatic islets and basically exerts effects that are opposite to those of insulin. It raises the blood glucose level by increasing the rate of glycogen breakdown (glycogenolysis) in the liver. It also promotes the formation of glucose from non-carbohydrate precursors (gluconeogenesis) in the liver. The primary stimulus for increased secretion of glucagon is a fall in the concentration of glucose in blood. During most types of exercise, the blood glucose concentration does not fall, but during prolonged exercise, when liver glycogen stores become depleted, a drop in the blood glucose concentration (hypoglycaemia) may occur.

Catecholamines promote liver glycogen breakdown and adipose tissue lipolysis

The catecholamines adrenaline and noradrenaline are released from the adrenal medulla. Noradrenaline is also released from sympathetic nerve endings and leakage from such synapses appears to be the main source of the noradrenaline found in blood plasma. The catecholamines have many systemic effects throughout the body including stimulation of the heart rate and contractility and alteration of blood vessel diameters. They also influence substrate availability, with the effects of adrenaline being the more important of the two. Adrenaline, like glucagon, promotes glycogenolysis in both liver and muscle. Adrenaline also promotes lipolysis in adipose tissue, increasing the availability of plasma FFA, and inhibits insulin secretion. The primary stimulus for catecholamine secretion is the activation of the sympathetic nervous system by stressors such as exercise, hypotension, and hypoglycaemia. Substantial increases in the plasma catecholamine concentration can occur within seconds of the onset of high-intensity exercise. However, the relative exercise intensity has to be above about 50% VO_{2max} to significantly elevate the plasma catecholamine concentration.

Growth hormone and cortisol are also involved in the control of fuel availability

Growth hormone, secreted from the anterior pituitary gland, also stimulates mobilization of FFA from adipose tissue, and increases in plasma growth hormone concentration are related to the intensity of exercise performed. During prolonged strenuous exercise cortisol secretion from the adrenal cortex is increased. Cortisol is a steroid hormone that increases the effectiveness of the actions of catecholamines in some tissues (e.g. its actions further promote lipolysis in adipose tissue). However, its main effects are to promote protein degradation and amino acid release from muscle and to stimulate gluconeogenesis in the liver. The primary stimulus to cortisol secretion is stress-induced release of adrenocorticotrophic hormone from the anterior pituitary gland.

 A possible role of cytokines in fuel mobilization

Recent evidence suggests that active muscles release some cytokines, including interleukin-6 (IL-6), during prolonged exercise. Cytokines are protein messenger molecules that act in a hormone-like manner, although until recently it was thought that they were almost exclusively produced by cells of the immune system. The word 'cytokine' derives from the Greek, 'cyto' meaning cell and 'kine' meaning movement. Cytokines are peptides or proteins typically defined as 'molecules that are produced and released by cells of the immune system and mediate the generation of immune responses'. This definition of cytokines does not tell the whole story, however. It is more accurate to state that 'cytokines are secreted molecules that may exert specific effects both on the cell from which they are secreted (autocrine effects) and on other cells (paracrine effects)'. The important distinction between these definitions is that while cytokines were initially identified as molecules released by cells of the immune system, their origins and influence spread far beyond that of the immune system alone. Indeed, exercise has proved to be a fascinating example of how cells of non-immune origin are able to produce and release specific cytokines. It has been suggested that muscle-derived cytokines and other peptides that exert paracrine or endocrine effects should be named 'myokines'.

Many recent studies have demonstrated an increase in the circulating concentration of several cytokines following prolonged strenuous exercise. Increases in the circulating concentrations of both pro-inflammatory cytokines (e.g. interleukin (IL)-1β, tumour necrosis factor (TNF)-α), and anti-inflammatory cytokines (e.g. IL-6 and IL-10), cytokine inhibitors (e.g. IL-1 receptor antagonist and soluble TNF receptors), chemokines (e.g. IL-8, macrophage inflammatory protein, and monocyte chemotactic protein-1), and colony-stimulating factors have been reported following endurance exercise. Skeletal muscle also has the capacity to produce other cytokines including IL-8 and IL-15 and contractile activity appears to play a role in regulating the expression of these cytokines in the muscle fibres. However, the finding of an increase in the circulating IL-6 concentration following prolonged exercise is the most marked, and consistent, exercise-induced response of any cytokine so far examined (See Figure 5.13).

Tracer methodology. A tracer is simply a marker that allows the fate of a compound in the body to be followed. Various tracer methodologies are available, but tracer methods involving the use of isotopes incorporated into compounds of interest are perhaps most common. These may be radioactive or stable isotopes; the latter are much preferred for health and safety reasons, but much of the early work involving metabolic tracers was carried out using radioactive tracers. There are convenient stable isotopes of hydrogen, carbon, oxygen, and nitrogen that can be used as tracers for a wide range of organic molecules. The isotope is incorporated into the compound of interest and this is introduced into the body: depending on the study, this may be ingested, injected, or infused. The fate of the labelled molecule can be followed by suitable sampling methods. For example, an infusion of glucose labelled with carbon-13 (^{13}C) may be given to a volunteer during exercise: the appearance of ^{13}C in expired CO_2 will give a measure of the rate of oxidation of blood glucose. Alternatively, the tracer might be added to a glucose-containing drink: in this case, the amount of label recovered in the breath as CO_2 is a measure of how much the ingested glucose is contributing to total carbohydrate oxidation. The latter can, of course, be quantified if the rate of oxygen uptake and the respiratory exchange ratio are known.

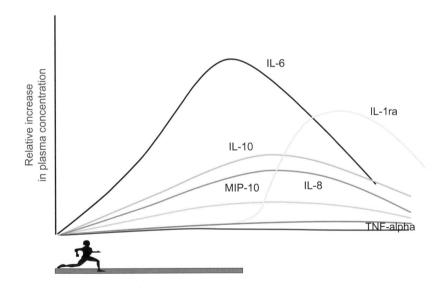

Figure 5.13

The cytokine response to prolonged strenuous exercise. From Febbraio and Pedersen (2002).

The main sources of IL-6 *in vivo* are activated monocytes/macrophages, fibroblasts, and endothelial cells. However, it has been shown that monocytes were unlikely to be the source of the exercise-induced increase in the plasma IL-6 concentration as a prolonged bout of strenuous exercise caused no changes in the amount of IL-6 mRNA or IL-6 protein detected in peripheral blood mononuclear cells despite an elevation in the plasma IL-6 level. Several possible sites of origin were suggested for the exercise-induced increase in the circulating level of IL-6, with contracting skeletal muscle receiving the most attention. Specifically, the intriguing finding that mRNA for IL-6 was elevated in the previously contracting skeletal muscle following prolonged exercise led to the hypothesis that strenuous exercise caused the destruction of contracting myofibres, triggering an inflammatory response and the subsequent release of IL-6 into the systemic circulation. However, it soon became apparent that muscle damage *per se* was only a minor contributor to the exercise-induced rise in the circulating level of IL-6 and that, in fact, it was normally contracting muscle fibres that were producing IL-6 and releasing it into the circulation.

In a study by Steensberg *et al.* (2000) from the Copenhagen Muscle Research Centre catheters were placed into the femoral artery of one leg and the femoral vein of both legs and subjects performed 5 h of single-legged knee extensor exercise, a purely concentric exercise model. The results of this study were intriguing and demonstrated that the contracting leg releases IL-6 and that this release almost exclusively accounts for the elevated systemic IL-6 concentration (Figure 5.14). Importantly, despite the same supply of various hormones, metabolites, and other potential mediators of IL-6 production from the femoral arteries to the resting and exercising legs, no release of IL-6 was detected from the resting leg. This demonstrates that IL-6 release from the muscle is absolutely dependent upon muscle contraction, and that secreted factors (e.g. adrenaline) do not play an important role in the exercise-induced increase in the systemic IL-6 concentration. Importantly, this study does not conclusively demonstrate that skeletal muscle is the source of the exercise-induced increase in the systemic IL-6 concentration as the arterio-venous difference technique is only able to measure the net uptake or release of a given molecule (in this case IL-6) over a specific region

Figure 5.14

Release of IL-6 from resting and exercising legs during 5 h of single-legged knee extensor exercise. From Steensberg *et al.* (2000).

of tissue (in this case the upper leg). Therefore, although this study provided strong evidence that IL-6 is released from contracting muscle, it is possible that other cellular sources within the upper leg, such as resident tissue macrophages, fibroblasts in connective tissue, the endothelium of the muscle capillary bed, adipose tissue, or bone, may have contributed to increased release of IL-6 from the upper leg during exercise. However, subsequent studies confirmed the presence of increased IL-6 protein accumulation and increased IL-6 gene expression in muscle fibres after prolonged exercise using immunohistochemistry (staining with fluorescence-labelled monoclonal antibody to IL-6) and *in situ* hybridization, respectively.

The release of IL-6 from the contracting muscles and subsequent accumulation in the systemic circulation is closely related to the duration of the exercise bout. It is well known that during prolonged exercise the level of muscle glycogen in the contracting skeletal muscles decreases, and it has been shown that during prolonged exercise, IL-6 is released from skeletal muscles in response to a reduction in muscle glycogen stores, within the contracting myofibres. As muscle glycogen levels decrease, the contracting muscles' reliance on blood glucose as a substrate for energy increases. Thus, IL-6 is released from the contracting muscles and acts as a signal to the liver to increase its glucose output and prevent a drastic, exercise-induced, fall in the blood glucose concentration as illustrated in Figure 5.15. This hypothesis is supported by the observation that carbohydrate ingestion during

Figure 5.15

Biological role of contraction-induced IL-6. In response to muscle contractions, both type I and type II skeletal muscle fibres express the myokine IL-6, which subsequently exerts its effects both locally within the muscle (e.g., through activation of AMP-kinase) and, when released into the circulation, peripherally in several organs in a hormone-like fashion. Specifically, in skeletal muscle, IL-6 acts in an autocrine or paracrine manner to increase glucose uptake and fat oxidation. IL-6 is also known to increase hepatic glucose production during prolonged exercise and lipolysis in adipose tissue.

exercise, which provides an exogenous source of glucose and helps to maintain the blood glucose concentration, attenuates the systemic IL-6 concentration.

Exercise-induced elevations of plasma IL-6 may also influence fat metabolism. When recombinant human IL-6 was infused into resting humans, increasing the plasma IL-6 concentration to levels observed during prolonged exercise, the rates of lipolysis and fat oxidation were increased. It has also been shown that IL-6-deficient mice develop mature-onset obesity and that when these mice were treated with IL-6 for 18 days, their body weight decreased. These studies identify IL-6 as a possible regulator of fat metabolism and support the hypothesis that contracting muscles release IL-6 in a hormone-like manner to increase substrate mobilization.

Strenuous exercise increases plasma concentrations of cortisol, glucagon, adrenaline, and noradrenaline. Infusion of rhIL-6 into resting humans to mimic the exercise-induced plasma levels of IL-6 increases plasma cortisol in a similar manner. In contrast, the same rhIL-6 infusion does not change plasma adrenaline, noradrenaline, or insulin levels in resting healthy young subjects. Therefore, muscle-derived IL-6 may be partly responsible for the cortisol response to exercise, whereas other hormonal changes cannot be ascribed to IL-6. Stimulation of cortisol secretion by IL-6 may be due to an effect of IL-6 on the hypothalamus, stimulating the release of ACTH from the anterior pituitary gland or by a direct effect of IL-6 on cortisol release from the adrenal glands; evidence for both mechanisms exists. In addition, it is known that relatively small increases in plasma levels of IL-6 induce the two anti-inflammatory cytokines IL-1ra and IL-10 together with C-Reactive Protein, and inhibit production of the pro-inflammatory cytokine TNF-α. According to the low-grade inflammation theory high plasma TNF-α concentrations have been suggested as an important mechanism in insulin resistance and arteriosclerosis, and regular exercise protects against these disorders. This exercise effect is likely partly to be mediated through the inhibitory effects of muscle-derived IL-6 on TNF-α production. In addition, other mediators such as adrenaline and cortisol that are both generally immunosuppressive may also contribute to the 'anti-inflammatory' effects of exercise.

Altogether, these findings suggest that muscle-derived IL-6 may play a role in regulating both fuel metabolism and the immune system during exercise. Manipulating exercise-induced IL-6 levels may provide a mechanism to explain how exercise either reduces the susceptibility to, or improves the symptoms of, diseases associated with low-grade inflammation such as type 2 diabetes, atherosclerosis, and possibly some autoimmune diseases. In addition, muscle-derived IL-6 may also reduce the inflammatory response in the exercised muscles and could even play a role in the development of central fatigue and the mood changes that accompany overtraining. Intriguingly, other tissues have been shown to contribute to the exercise-induced increase in the systemic IL-6 concentration in addition to the contracting skeletal muscle. For example, the brain has been shown to release IL-6 during exercise, and IL-6 gene expression within adipose tissue has been shown to be increased in response to exercise. However, the stimulus for the increased production and release of IL-6 from these tissues is unclear, and whether IL-6 released from the brain and adipose tissue during exercise plays a similar or different role to muscle derived IL-6 is, at present, not known.

Fatigue in prolonged exercise

Eventually, ATP production is compromised as muscle and hepatic carbohydrate stores are depleted

The term prolonged exercise is usually used to describe exercise intensities that can be sustained for 30–180 min. As the rate of ATP demand is relatively low compared with high-intensity exercise, PCr, carbohydrate, and fat can all contribute to energy production. The rates of PCr degradation and lactate production during the first minutes of prolonged exercise are closely related to the intensity of exercise performed, and it is likely that energy production during this period would be compromised without this contribution from anaerobic metabolism. However, once a steady state has been reached, carbohydrate and fat oxidation become the principal means of resynthesizing ATP. Muscle glycogen is the principal fuel during the first 30 min of exercise at 60–80% VO_{2max}. During the early stages of exercise, fat oxidation is limited by the delay in the mobilization of fatty acids from adipose tissue. At rest, following an overnight fast, the plasma FFA concentration is about 0.4 mmol/l. This is commonly observed to fall during the first hour of moderate intensity exercise, followed by a progressive increase as lipolysis is stimulated by the actions of catecholamines, glucagon, and cortisol. During very prolonged exercise, the plasma FFA concentration can reach 1.5–2.0 mmol/l and muscle uptake of blood-borne FFA is proportional to the plasma FFA concentration. The glycerol released from adipose tissue cannot be used directly by muscle, which lacks the enzyme glycerol kinase. However, glycerol (together with alanine and lactate) is taken up by the liver and used as a gluconeogenic precursor to help maintain liver glucose output as liver glycogen levels decline. The utilization of blood glucose is greater at higher work rates and increases with exercise duration during prolonged submaximal exercise and peaks after about 90 min (as illustrated in Figure 5.16). The decline in blood glucose uptake after this time is attributable to the increasing availability of plasma FFA as fuel (which appears to directly inhibit muscle glucose uptake) and the depletion of liver glycogen stores.

The rate of ATP resynthesis from fat oxidation alone cannot meet the ATP requirement for exercise intensities higher than about 50–60% VO_{2max}

At marathon running pace, muscle carbohydrate stores alone could fuel about 80 min of exercise before becoming depleted (Table 5.3). However, the simultaneous utilization of body fat and hepatic carbohydrate stores enables ATP

Figure 5.16

The utilization of blood glucose is greater at higher power outputs and increases with exercise duration during prolonged submaximal exercise and at 60% VO$_{2max}$ it peaks after about 90 min. The decline in blood glucose uptake after this time is attributable to the increasing availability of plasma FFA as fuel (which appears to directly inhibit muscle glucose uptake) and the depletion of liver glycogen stores.

Table 5.3 Body energy stores in a typical 70-kg man and the time that these would last for if used as the sole fuel for continuous exercise at marathon running pace (assuming an energy expenditure of 80 kJ/min).

	Amount available (g)	Time to depletion (min)
Muscle glycogen	300	60
Liver glycogen	100	20
ECF glucose	20	4
Adipose triglyceride	12 000	5550
Muscle triglyceride	300	140
Protein	10 000	2250

production to be maintained and exercise to continue. Eventually, however, ATP production becomes compromised as muscle and hepatic carbohydrate stores are depleted and fat oxidation is unable to increase sufficiently to offset this deficit. The rate of ATP resynthesis from fat oxidation alone cannot meet the ATP requirement for exercise intensities higher than about 50–60% VO$_{2max}$. It is currently unknown which factor limits the maximal rate of fat oxidation during exercise (i.e. why it cannot increase to compensate for carbohydrate depletion), but it must precede acetyl-CoA formation, as from this point fat and carbohydrate share the same fate. The limitation may reside in the rate of uptake of FFA into muscle from blood or the transport of FFA into the mitochondria rather than in the rate of β-oxidation of FFA in the mitochondria.

It is generally accepted that the glucose–fatty acid cycle regulates the integration of carbohydrate and fat oxidation during prolonged exercise. However, although this may be true of resting muscle, Dyke *et al.* (1993) have suggested that the cycle does not operate in exercising muscle and that the site of regulation must reside elsewhere (e.g. at the level of phosphorylase and/or malonyl-CoA).

From the literature, it would appear that the integration of muscle carbohydrate and fat utilization during prolonged exercise is complex and unresolved.

> Endurance performance is strongly related to pre-exercise muscle glycogen content

The glycogen store of human muscle is fairly insensitive to change in sedentary individuals. However, the combination of exercise and dietary manipulation can have dramatic effects on muscle glycogen storage. A clear positive relationship has been shown to exist between muscle glycogen content and subsequent endurance performance. Furthermore, the ingestion of carbohydrate during prolonged exercise is known to decrease muscle glycogen utilization and fat mobilization and oxidation, and to increase the rate of carbohydrate oxidation and endurance capacity. It is clear, therefore, that the contribution of orally ingested carbohydrate to total ATP production under these conditions must be greater than that normally derived from fat oxidation. The precise biochemical mechanism by which muscle glycogen depletion results in fatigue is presently unresolved. However, it is plausible that the inability of muscle to maintain the rate of ATP synthesis in the glycogen depleted state results in ADP and P_i accumulation and consequently fatigue development.

> The development of hypoglycaemia may contribute to fatigue in prolonged exercise

Unlike skeletal muscle, starvation rapidly depletes the liver of carbohydrate. The rate of hepatic glucose release in resting post-absorptive individuals is sufficient to match the carbohydrate demands of only the central nervous system. Approximately 70% of this release is derived from liver carbohydrate stores and the remainder from liver gluconeogenesis. During exercise, the rate of hepatic glucose release has been shown to be related to exercise intensity. Ninety per cent of this release is derived from liver carbohydrate stores, ultimately resulting in liver glycogen depletion. Thus, carbohydrate ingestion during exercise could also delay fatigue development by slowing the rate of liver glycogen depletion and helping to maintain the blood glucose concentration. Central fatigue is a possibility during prolonged exercise and undoubtedly the development of hypoglycaemia could contribute to this.

The role of the brain in fatigue

The nature of fatigue has been investigated for well over a century. The early observations were largely descriptive as the techniques necessary to establish the underlying mechanisms were not available. Nonetheless, the early physiologists who conducted studies in this area were able to draw meaningful conclusions from their observations. At the end of the nineteenth century, there was considerable debate as to whether fatigue actually occurred within the muscles themselves or whether it was a nervous phenomenon. To investigate this, Mosso devised an instrument to study fatigue in repeated contractions of the finger: he conducted many studies to look at the effects of mental work—as in delivering a lecture to a class of students—on fatigue of the muscles, and his measurements led him to conclude that both the nervous system and the muscles were involved in the fatigue process. This was expressed rather elegantly by Bainbridge who wrote, in 1919, that "There appear to be two types of fatigue, one arising entirely within the central nervous system, the other in which fatigue of the muscles themselves is superadded to that of the nervous system".

This description, apt though it seems, provided no indication of possible mechanisms. More recently, there has been much attention paid to the potential role of a 'central governor' that limits muscle performance by regulating the recruitment of muscle fibres. One possible input into this system might be the availability of carbohydrate fuel, with hypoglycaemia affecting brain function during prolonged exercise when liver glycogen stores are low. A reduced availability of carbohydrate substrate for the brain may be a contributing factor in some situations, but fatigue commonly occurs when blood glucose concentration appears to remain adequate. Although both liver and muscle glycogen stores may be very low at the end of exercise that lasts more than 2–3 h and is carried out in a cool or temperate environment, the fatigue that results from prolonged exercise in the heat seems not to be related to carbohydrate depletion.

The physical mechanisms involved in the 'central governor' are not well understood, but this is likely to have a neurological basis, which ultimately means that there must be a neurochemical mechanism, or more likely a number of mechanisms acting in concert. Various pharmacological interventions have been applied to test this hypothesis, and the outcomes are generally consistent with a role for some key central neurotransmitters, specifically dopamine, serotonin, and noradrenaline, in the fatigue process. Perhaps the most convincing evidence for the role of the brain in the fatigue process comes from pharmacological interventions. Amphetamines, which act on central dopamine (DA) receptors, are well known to enhance performance in exercise and are prohibited under the rules of the World Anti-Doping Agency (WADA). Various studies have demonstrated a marked increase in exercise capacity following the administration of amphetamines to both rodents and humans. Amphetamines are thought to enhance exercise performance through the maintenance of DA release late in exercise, as an elevation in catecholaminergic neurotransmission is typically linked to arousal, motivation, and reward. Changes in regional DA metabolism have also been implicated in the control of locomotion and posture in moving animals, which also may be an important aspect of the role of DA in fatigue.

Many different psychotropic drugs that act through changes to central dopaminergic and noradrenergic neurotransmission with varying degrees of receptor specificity to treat/manage psychiatric disorders are now available. Bupropion, which acts on central dopamine and noradrenaline receptors, has recently been shown to enhance endurance performance in the heat, but this effect was not seen when exercise was undertaken in temperate conditions. Even though a higher power output was maintained during exercise in the heat after administration of this drug, the subjects' perception of effort and thermal discomfort was the same on both the treatment and placebo trials. Core (rectal) temperature at the end of the trial was, however, higher on the treatment trials than on the placebo trials, which is consistent with the higher mean power output. It is possible that dopaminergic drugs may dampen or override inhibitory signals arising from the CNS to cease exercise due to hyperthermia, and enable an individual to maintain a high power output in spite of an elevated core temperature.

Views of the role of neurotransmission in the development of fatigue during prolonged exercise have centred around the neurotransmitter serotonin (5-hydroxytryptamine, 5-HT). Professor Eric Newsholme and colleagues suggested that changes in substrate mobilization occurring during prolonged exercise would result in an increased uptake of the amino acid tryptophan into the central nervous system. As serotonin synthesis is largely dependent up on the availability of tryptophan, they postulated that central serotoninergic synthesis, and therefore activity, would increase as a consequence of the metabolic changes taking place as exercise continued. Serotoninergic neurotransmission has been linked to feelings of lethargy and tiredness, and Newsholme proposed that this response could contribute to the fatigue process. The evidence for drugs acting on serotonin receptors is perhaps less convincing than the work on dopamine, with some studies showing a reduction in exercise capacity following the administration of 5-HT agonists, while others have shown no effect. An attraction of this hypothesis was the possibility to alter central 5-HT activity through the provision of nutritional supplements, including branched-chain amino acids, tryptophan, and carbohydrate. Attempts to influence performance by nutritional interventions designed to manipulate the availability of neurotransmitter precursors have, however, generally been unsuccessful, though there are some exceptions.

To further emphasize the potential involvement of dopaminergic neurones in the fatigue process, there is some recent evidence that rats selectively bred for either high or low running capacity show differences in the expression of genes related to dopaminergic function in some regions of the brain. Attempts to find a genetic basis for the success of specific athletic populations, such as the East African distance runners, have generally been unsuccessful, perhaps because they have looked in the wrong place.

Nutrition and endurance exercise performance

Diet affects endurance exercise performance

The diet during training can influence the intensity and duration of training sessions that can be sustained by an athlete. Furthermore, nutrition in the days leading up to competition and on the day of competition (including what is consumed during a race) can have a marked influence on endurance performance.

Carbohydrate and protein

Glycogen availability is an important determinant of endurance performance

Key nutritional challenges for the endurance athlete are to ensure a sufficient energy intake to meet the high energy demands of training and competition and to ensure an adequate intake of carbohydrate. The energy demands of intensive training are high and endurance athletes may need to consume more than twice as much energy as their sedentary counterparts. Because of the demands on carbohydrate as a fuel in daily training, the diet may need to supply 6–10 g of carbohydrate per kg of body mass per day: this may be twice the total storage capacity of the liver and muscle glycogen combined. It is hardly surprising, therefore, that endurance athletes seem to be constantly eating high-carbohydrate snacks.

Eating carbohydrate-containing foods soon after training seems to be important for rapid restoration of muscle glycogen and athletes are encouraged to eat about 50–100 g of carbohydrate as soon as possible after training. In the light of the emerging evidence about the role of protein intake in supplying essential amino acids for remodelling of the muscle tissues to take place (discussed in Chapter 8), it might be sensible to include some protein in these post-training snacks.

Among the first studies to employ the needle biopsy technique to study skeletal muscle metabolism in humans was a landmark study from Jonas Bergstrom and Eric Hultman that was published in *Nature* in 1966. This study involved only two subjects (the authors themselves) and they exercised by pedalling with one leg each by sitting on either side of a cycle ergometer. After they had gone as long as possible, they took muscle samples from each other, one from the exercised leg and one from the rested leg. As you can see in Figure 5.17, in both individuals the glycogen content of the exercised leg was reduced to almost zero, but the rested leg was unaffected by the exercise performed by the other leg. They then repeated the muscle sampling on both legs at 1, 2, and 3 days later. The glycogen content of the rested leg remained more or less unchanged, but that of the exercised leg increased progressively to levels far above those normally seen. They later showed that feeding a high-carbohydrate diet was necessary to achieve this effect: feeding a low-carbohydrate diet after exercise prevented restoration of the muscle glycogen stores. They also showed that endurance capacity in cycling at about 70% of VO_{2max} was closely related to the initial muscle glycogen levels: the higher the glycogen content, the faster it was used but the longer it lasted before reaching critically low levels (see Figure 5.18).

Figure 5.17

Changes in muscle glycogen content in two subjects after prolonged one-leg cycling exercise. There is a marked decrease in muscle glycogen content of the exercised leg during exercise and, although both legs are exposed to the same nutrient supply, only the exercised leg responds with a supercompensation of glycogen content. Redrawn from the data of Bergstrom and Hultman (1966).

These observations were published in the scientific literature in 1966 and 1967, but it was 1969 before they were applied to sport. The English marathon runner Ron Hill used the carbohydrate-loading diet in his preparations for the European Championship marathon held in Athens in 1969. One week before the race, he went for a long run to deplete muscle glycogen, and then ate a very low-carbohydrate diet for the next 3 days while continuing to train. He then reduced the length and speed of his daily runs and switched to a high-carbohydrate diet for the last 3 days before racing, so that he started the race with very high muscle glycogen stores. In the race itself, he was lying second, well behind the leader, at 35 km, but he ran strongly towards the end to win comfortably. After this, the 'carbo-loading' diet became popular with distance runners and pasta parties became a feature of pre-race preparations. It is now known that it is not necessary

Figure 5.18

It was established in the 1960s that the exercise capacity in prolonged exercise was proportional to the pre-exercise muscle glycogen concentration. When this was manipulated by altering the dietary carbohydrate intake, large changes in exercise performance resulted.

for well-trained athletes to follow such an extreme routine. Complete glycogen depletion is not necessary and neither is the low-carbohydrate diet phase. It is enough just to reduce the training load and increase the amount of carbohydrate in the diet for the last few days before racing, and this remains standard practice for most elite endurance athletes.

As mentioned previously, the ingestion of carbohydrate during prolonged exercise has been shown to decrease muscle glycogen utilization and fat mobilization and oxidation, and to increase the rate of carbohydrate oxidation and endurance capacity. Carbohydrate ingestion during exercise could also delay fatigue development by slowing the rate of liver glycogen depletion and helping to maintain the blood glucose concentration. Improvements in endurance capacity or performance are generally seen when carbohydrate is ingested at a rate of 30–60 g/h and when the exercise duration is an hour or more.

Caffeine and carnitine

Caffeine improves endurance exercise performance

If glycogen depletion is the primary cause of fatigue, an alternative to increasing the amount of glycogen stored in the muscles is to increase the contribution of fatty acid oxidation to energy supply, thus reducing the rate of glycogen depletion. Many different nutritional strategies are used by endurance athletes in an attempt to spare muscle glycogen and increase the reliance on fat as a metabolic substrate. These include the use of high-fat diets in training and the use of a range of nutritional supplements, including caffeine, carnitine, and other purported 'fat burners'.

Caffeine is a trimethylxanthine compound and is not produced by the human body, although it can be considered as a nutritional ergogenic aid because it is a natural constituent of several common beverages, particularly coffee. Taken in relatively small amounts prior to exercise, caffeine can improve performance in both brief intense effort and endurance exercise. The evidence for beneficial effects of caffeine on performance seems strong, and this is usually attributed to its stimulation of adrenaline release from the adrenal glands and the subsequent stimulation of lipolysis, which increases the mobilization of fatty acids, thus resulting in a glycogen-sparing effect. In exercise that can only be sustained for a few minutes, however, glycogen depletion is not the cause of fatigue, but performance can be improved by caffeine ingestion, so other mechanisms must also be operating. It is now thought, however, that other effects of caffeine, including effects on both the central nervous system and the muscles themselves, may be at least partially responsible for the improvements in performance that are observed. Possible mechanisms involve stimulatory effects on the brain by an action on adenosine receptors (thereby increasing mental arousal and modifying the perception of effort) and, in muscle, facilitating the release of calcium from its storage sites in the muscle fibres, enabling calcium to stimulate muscle contraction more effectively and thus improve force generation.

Well-controlled scientific studies have established that caffeine is an effective ergogenic aid in a variety of exercise modes (e.g. running and cycling) and for a wide range of exercise intensities. Thus, participants in many sports could potentially benefit from its performance-enhancing properties. For runners, caffeine appears to be particularly effective for middle-distance and endurance events.

Caffeine use in competition was previously prohibited under IOC doping regulations in amounts that give a urinary caffeine concentration of 12 mg/l or more. However, following a review the unrestricted use of caffeine is now permitted, although its use in competition is monitored by WADA. The effects on both exercise performance and cognitive function can be seen at doses of about 1–3 mg/kg body mass, but tolerance varies between individuals, so caution is necessary. The amounts that have been tested in research studies that have investigated caffeine's ergogenic effects on endurance exercise performance have ranged from 1 to 15 mg/kg body mass. For a 70-kg subject this represents a total dosage of 70–1050 mg of caffeine. A typical cup of brewed coffee contains about 100 mg of caffeine. The purported performance-boosting effects of some herbal supplements such as guarana may be attributed to their high caffeine content. Some cola drinks contain caffeine, typically about 40 mg per 400 ml can.

> There is limited evidence that carnitine is an effective ergogenic aid

On the basis of its role in the transport of fatty acids into mitochondria for oxidation, carnitine is widely promoted for use by endurance athletes and as a weight (fat)-loss product. The theory is sound, but until recently there was little experimental evidence to support this. Some animal studies suggested that supplementation of the diet with carnitine may increase fat oxidation during exercise and improve endurance exercise performance, but most human studies of carnitine administration showed no effect on exercise performance and no increased reliance on fat as an energy substrate. Studies that have looked at carnitine levels in human muscle generally show that this is not affected by supplementation, so the lack of effect is perhaps not surprising. More recently, however, it has been found that intravenous infusion of carnitine along with simultaneous infusion of insulin to maintain high levels in the circulation can stimulate the uptake of carnitine into skeletal muscle. Simply increasing the carnitine concentration on its own had no effect. Ingesting a large amount of carbohydrate along with carnitine appears to have a similar effect, again due to a stimulatory action of insulin on the membrane carnitine transporter. This effect seems to be via an increase in the number of active carnitine transporters in the muscle membrane. Increasing muscle total carnitine content in resting healthy humans does appear to reduce muscle glycolysis and increase fat oxidation. The potential for improving exercise performance and for promoting weight loss remains to be fully explored, but the early results look promising.

KEY POINTS

1. The term prolonged exercise is usually used to describe exercise intensities that can be sustained for 30–180 min. As the rate of ATP demand is relatively low compared with high-intensity exercise, PCr, CHO, and fat can all contribute to energy production.

2. The rates of PCr degradation and lactate production during the first minutes of prolonged exercise are closely related to the intensity of exercise performed, and it is likely that energy production during this period would be compromised without this contribution from anaerobic metabolism. However, once a steady state has been reached, CHO and fat oxidation become the principal substrates.

3. Under normal conditions, muscle CHO stores alone could fuel about 80 min of exercise before becoming depleted. However, the simultaneous utilization of body fat and hepatic CHO stores enables ATP production to be maintained and exercise to continue. Ultimately, however, ATP production becomes compromised as muscle and hepatic CHO stores become depleted and fat oxidation is unable to increase to offset this deficit.

4. Gluconeogenesis—the synthesis of carbohydrate from non-carbohydrate sources—occurs primarily in the liver, and can help to maintain the blood glucose concentration that is important not only for exercising muscle but also for tissues such as the brain and red blood cells, which are almost entirely dependent on the availability of glucose.

5. It is currently unknown which factor limits the maximal rate of fat oxidation during exercise (that is why it cannot increase to compensate for CHO depletion), but it must precede acetyl-CoA formation, as from this point fat and carbohydrate share the same fate. The energy source for muscle contraction is ATP, which is continuously regenerated during exercise from PCr, anaerobic metabolism of glycogen or glucose, or aerobic metabolism of acetyl-CoA derived principally from breakdown of carbohydrate or fat.

6. The TCA cycle and oxidative phosphorylation occur in the mitochondria. In the aerobic resynthesis of ATP, the primary role of oxygen is to act as the final electron acceptor in the electron-transport chain and to combine with hydrogen to form water.

7. The principal storage form of fat in the body is triacylglycerol, most of which is located in white adipose tissue. Triacylglycerol stores are also found in liver and muscle and as lipoproteins in blood.

8. Triacylglycerols are formed by the sequential linking of three fatty acid molecules to glycerol. The synthesis of fatty acids from acetyl-CoA requires ATP and NADPH and occurs in the cytoplasm of the cells of the liver and adipose tissue.

9. Muscles cannot oxidize triacylglycerols directly. The triacylglycerol molecule must first be broken down into its fatty acid and glycerol components in the process called lipolysis. The latter is catalysed by a hormone-sensitive lipase found in adipocytes and muscle fibres. Lipoprotein lipase in the capillary endothelium breaks down plasma triacylglycerols.

10. The principal sources of fat fuels for exercise are blood-borne FFA derived from adipose tissue and intramuscular triacylglycerol. The uptake of FFA into muscle is a carrier-mediated process that exhibits saturation kinetics.

11. Fatty acids undergo β-oxidation in the mitochondria, yielding acetyl-CoA, NADH, and $FADH_2$. Acetyl-CoA can enter the TCA cycle and the reduced coenzymes pass their electrons and hydrogen to oxygen via the mitochondrial respiratory chain. Hence, utilization of fat energy requires oxygen.

12. Lipolysis is activated during exercise via the actions of adrenaline and glucagon.

13. Fat oxidation makes an increasing contribution to ATP regeneration as exercise duration increases. In exercise lasting several hours, fat may supply almost 80% of the total energy required.

14. Fat oxidation can only supply ATP at a maximum rate of about 1 mmol/s/kgdm, equivalent to the requirement when exercising at an intensity of about 50–60% VO_{2max}. It is not possible for fat to supply ATP at the rate required at higher exercise intensities. The principal limitation may be the rate of entry of FFA into the mitochondrion.

15. The rate of FFA oxidation in muscle is related to the plasma FFA concentration and blood flow, and is also regulated in part by the oxidative capacity of the recruited muscle fibres and the availability of carbohydrate stores.

16. Endurance training adaptations increase the capacity of muscle to oxidize fat.

17. Muscle can only oxidize 6 of the amino acids to any significant extent. Amino acid oxidation has been estimated to contribute up to only about 15% to energy expenditure in resting conditions. During exercise, this relative contribution likely decreases because of an increasing importance of carbohydrate and fat as fuels.

During very prolonged exercise, when carbohydrate availability becomes limited, amino acid oxidation may increase somewhat but the contribution of amino acid oxidation to energy expenditure still only amounts to a maximum of about 10% of total energy expenditure.

18. It is generally accepted that the glucose-fatty acid cycle regulates the integration of CHO and fat oxidation during prolonged exercise. However, while this may be true of resting muscle, evidence suggests that the cycle does not operate in exercising muscle and that the site of regulation must reside elsewhere (e.g. at the level of phosphorylase and/or malonyl-CoA). From the literature it would appear that the integration of muscle CHO and fat utilization during prolonged exercise is complex and unresolved.

19. The CHO store of human muscle is fairly insensitive to change in sedentary individuals. However, the combination of exercise and dietary manipulation can have dramatic effects on subsequent muscle CHO storage. Furthermore, a clear positive relationship has been shown to exist between muscle CHO content and subsequent prolonged exercise performance.

20. Unlike skeletal muscle, starvation rapidly depletes the liver of CHO. The rate of hepatic glucose release in resting post-absorptive individuals is sufficient to match the CHO demands of only the central nervous system. Approximately 70% of this release is derived from liver CHO stores and the remainder from liver gluconeogenesis.

21. During exercise, the rate of hepatic glucose release has been shown to be related to exercise intensity. Ninety per cent of this release is derived from liver CHO stores, ultimately resulting in liver glycogen depletion. The exact mechanisms responsible for the regulation of hepatic glucose release during exercise are unresolved. Hepatic glucose uptake following exercise has been shown to be at least partly dependent on the form of CHO presented to the liver.

22. The improvement in exercise capacity following endurance training has been interpreted by some to be due to an increase in fat oxidation reducing muscle

KEY POINTS *(CONTINUED)*

glycogen utilization. This has led to the hypothesis that fat ingestion prior to exercise will improve endurance capacity. However, the maximal rate of fat oxidation is insufficient to match the rate of ATP resynthesis required during prolonged exercise at >60% VO_{2max} and, as might be expected therefore, performance has usually been found to be impaired under these conditions.

23. The exact biochemical mechanism by which muscle CHO depletion results in fatigue is presently unresolved. However, it is plausible that the inability of muscle to maintain the rate of ADP rephosphorylation in the glycogen-depleted state results in ADP and P_i accumulation and consequently fatigue development.

24. Several factors influence the type of substrate used to fuel muscular work. These include: substrate availability, diet, intensity and duration of exercise, training status, hormones, cytokines, prior exercise, and environmental conditions.

25. A number of hormones are involved in the integration and control of energy metabolism, including especially insulin, which promotes carbohydrate and fat storage, and glucagon, whose actions are generally antagonistic to those of insulin. Adrenaline and noradrenaline stimulate lipolysis as well as carbohydrate mobilization and metabolism at times of stress. Lipolysis may also be stimulated by the actions of IL-6, which is released from muscle in increasing amounts during prolonged exercise.

26. Exercise results in an increase in the circulating and tissue levels of numerous cytokines. Release of IL-6 from contracting muscle appears to be the main source of the elevated plasma IL-6 concentration during exercise. The brain also releases IL-6 during exercise, whereas there is a net uptake of circulating IL-6 by the liver. Muscle damage is not primarily responsible for the elevated concentration of cytokines in the plasma during exercise. However, the production and/or

release of some cytokines (notably IL-6) are increased when muscle glycogen content is depleted.

27. IL-6 appears to act in a hormone-like manner and is involved in increasing substrate mobilization (release of glucose from the liver and fatty acids from adipose tissue) during prolonged exercise. IL-6 also induces secretion of cortisol, IL-1ra, IL-10, and C-Reactive Protein and so has generally anti-inflammatory effects.

28. The ingestion of CHO during the hours before exercise ensures that hepatic CHO stores are optimal. There is no reason to believe that such a procedure impairs exercise performance as a result of the insulin rebound effect. Indeed, it is now generally accepted that CHO ingestion prior to prolonged exercise will improve performance.

29. The ingestion of CHO during prolonged exercise has been shown to decrease muscle glycogen utilization and fat mobilization and oxidation, and to increase CHO oxidation and endurance capacity during prolonged exercise. It is clear, therefore, that its contribution to total energy production under these conditions must be greater than that normally derived from fat oxidation.. Evidence suggests that CHO ingestion during prolonged exercise exerts its main functional and metabolic effects on Type-I muscle fibres. CHO ingestion could also retard fatigue development by delaying the rate of liver glycogen depletion and, thereby, the development of hypoglycaemia during exercise.

30. Caffeine is an effective ergogenic aid and there are several plausible mechanisms to explain its performance-enhancing effects. Some animal studies suggest that supplementation of the diet with carnitine may increase fat oxidation during exercise and improve endurance exercise performance, but most human studies have not reported a beneficial effect on metabolism or performance, which has been attributed to a failure to elevate muscle carnitine content by dietary supplementation in humans.

Selected further reading

Bangsbo J (1997). Physiology of muscle fatigue during intense exercise. In: *The clinical pharmacology of sport and exercise* (ed. Reilly T and Orme M). Amsterdam: Elsevier.

Bergstrom J and Hultman E (1966). Muscle glycogen synthesis after exercise: an enhancing factor localized to the muscle cells in man. *Nature* 210: 309–310.

Bergstrom J *et al.* (1967). Diet, muscle glycogen and physical performance. *Acta Physiologica Scandinavica* 71: 140–150.

Dyke DJ *et al.* (1993). Regulation of fat-carbohydrate interaction in skeletal muscle during intense aerobic cycling. *American Journal of Physiology* 265: E852–E859.

Galbo H (1983). *Hormonal and metabolic adaptation to exercise.* New York: Verlag.

Graham TE (2001). Caffeine, coffee and ephedrine: impact on exercise performance and metabolism. *Canadian Journal of Applied Physiology* 26: S103–S119.

Green HJ (1991). How important is endogenous muscle glycogen to fatigue in prolonged exercise? *Canadian Journal of Physiology and Pharmacology* 69: 290–297.

Febbraio MA and Pedersen BK (2002). Muscle-derived interleukin-6: mechanisms for activation and possible biological roles. *FASEB Journal* 16: 1335–1347.

Hargreaves M and Spriet L (ed.) (2006). *Exercise metabolism.* 2nd edn. Champaign, IL: Human Kinetics.

Hargreaves M *et al.* (2004). Pre-exercise carbohydrate and fat metabolism: effects on metabolism and performance. *Journal of Sports Sciences* 22: 31–38.

Maughan RJ (1999). Nutritional ergogenic aids and exercise performance. *Nutrition Research Reviews* 12: 255–280.

Maughan RJ *et al.* (2007). Exercise, heat, hydration and the brain. *Journal of the American College of Nutrition* 26: 604S–612S.

Pedersen BK and Febbraio MA (2008). Muscle as an endocrine organ: focus on muscle-derived interleukin-6. *Physiological Reviews* 88: 1379–1406.

Pedersen BK *et al.* (2007). Role of myokines in exercise and metabolism. *Journal of Applied Physiology* 103: 1093–1098.

Randle PJ *et al.* (1963). The glucose fatty acid cycle: its role in insulin sensitivity and the metabolic disturbances of diabetes mellitus. *Lancet* 1: 785–789.

Steensberg A *et al.* (2000). Production of interleukin-6 in contracting human skeletal muscles can account for the exercise-induced increase in plasma interleukin-6. *Journal of Physiology* 529: 237–242.

Stephens FB *et al.* (2006). An acute increase in skeletal muscle carnitine content alters fuel metabolism in resting human skeletal muscle, *Journal of Clinical Endocrinology and Metabolism* 91: 5013–5018.

Wagenmakers AJ (1998). Protein and amino acid metabolism in human muscle. *Advances in Experimental Medicine and Biology* 441: 307–319.

Wallenius V *et al.* (2002). Interleukin-6-deficient mice develop mature-onset obesity. *Nature Medicine* 8: 75–79.

Watson P *et al.* (2005). Blood-brain barrier integrity may be threatened by exercise in a warm environment. *American Journal of Physiology* 288: R1689–1694.

Williams MH (1998). *The ergogenics edge.* Champaign, IL: Human Kinetics.

Zammitt VA *et al.* (2009). Carnitine, mitochondrial function and therapy. *Advanced Drug Delivery Reviews* 61: 1353–1362.

6   The games player

Learning objectives

After studying this chapter, you should be able to...

1. appreciate the activity patterns and work rate in team games such as soccer;

2. describe the metabolic response to repeated bouts of high-intensity exercise with short recovery periods;

3. appreciate the relative importance of PCr and glycogen breakdown in supplying energy for ATP resynthesis in multiple-sprint activities;

4. describe the factors influencing the rate of resynthesis of PCr during recovery periods;

5. explain why the rate of muscle lactate accumulation declines with repeated sprints;

6. discuss mechanisms of fatigue in intermittent high-intensity exercise situations;

7. appreciate the nutritional needs of the games player;

8. discuss the possible benefits of dietary creatine supplementation for the games player.

Introduction

Multiple sprints with limited recovery are a characteristic of many team games and racquet sports

Games such as football, soccer, rugby, hockey, and tennis involve repetitive bouts of high-intensity exercise. Maintaining performance for the full duration of matches is obviously important to success and depends to a large degree on the ability of muscle to recover from the last exercise bout. The extent of recovery is affected by the intensity and duration of the last bout, the biochemical and physiological characteristics of the individual, and the time available for recovery before another bout of exercise has to be performed. In these sorts of games, the participants perform numerous sprints interspersed with variable periods of exercise of lower intensity (e.g. cruising, jogging, walking) or rest. The cycles of activity and rest are largely unpredictable as they are imposed by the pattern of play and vary greatly from player to player and from one match to another. Most team games last 70–90 min with a 10–15-min half-time break. In tennis it is not uncommon for men's matches to last for more than 3 h, although 30-s breaks are taken at change of ends after every odd-numbered game. Compared with continuous exercise activities such as running and cycling, relatively little attention has been directed to the metabolic responses to games such as soccer, rugby, hockey, or tennis, probably because of the lack of adequate experimental models to study these games in the laboratory. However, some standardized models of intermittent exercise have been developed in recent years that have gone some way to simulating the activity patterns observed in team games. This chapter describes how these protocols, as well as measurements made during competition itself, have shed some light on the demands of participation on the skeletal muscle recovery processes after intense exercise and their importance for metabolism and performance during one or more subsequent exercise bouts.

Activity patterns and work rate in games play

The total distance covered by outfield players varies between 8 and 13 km and the average player expends about 5000 kJ during a game

Most research attention in relation to team games has been directed to soccer, reflecting the universal popularity of this game. Soccer matches last 90 min (with a 15-min break at half-time) and each player performs about 1000 discrete bouts of action incorporating frequent changes of pace and direction as well as the execution of game skills such as passing, dribbling, and shooting. In English League matches players change activity every 6 s on average, and have brief rest periods averaging only 3 s every 2 min. Sprint distances average about 15 m and occur about every 90 s.

The total distance covered by outfield players varies between 8 and 13 km, with less than 4% of this distance covered while in possession of the ball. The distance covered in a match varies with positional roles in the team, with the greatest distances being covered by midfield players. On average, the total distance covered by outfield players during a match consists of about 37% jogging, 25% walking, 20% cruising (namely running faster than jogging, but not all-out sprinting), 11% sprinting, and 7% moving backwards. Heart rates of players have been recorded during friendly and competitive matches and typically have been found to be between 160 and 170 beats/min for the majority of the playing time. Although measurement of oxygen uptake is likely to interfere with normal play, some measurements have been made with players wearing portable respirometers during casual games. These studies indicate that energy expenditure is typically about 25–60 kJ/min and that the average player expends about 5000 kJ during a game (i.e. an average energy expenditure of about 55 kJ/min or an oxygen uptake of approximately 35 ml/kg/min). These values are likely to be underestimates of the true energy demands of competitive soccer, as the apparatus used to collect expired gas during play is likely to have hampered the activity of the players.

Players exercise at a relative intensity close to 75% VO_{2max} during a game

From measurements of heart rates of professional soccer players during match play it has been estimated that the energy expenditure is about 70 kJ/min and that the players are exercising at a relative intensity close to 75% VO_{2max}. This is not far from the intensity at which marathon runners race. Obviously, the soccer player covers considerably less distance compared with a marathon runner in 90 min and the physical demands of soccer are likely to be grossly underestimated if based on the distance covered alone. The intermittent nature of the exercise performed during a game means that players rely on anaerobic metabolism for brief bursts of sprinting. Jumping and kicking actions also rely on anaerobic energy supply. Blood lactate concentrations measured in professional players at half-time and at the end of a game are typically 4–6 mmol/l. It is likely that peak blood lactate concentrations of around 8–11 mmol/l occur during the course of a game and values above 12 mmol/l are occasionally observed.

The development of fatigue during a game seems to be related to depletion of the muscle glycogen stores

At high standards of competitive play it is noticeable that players suffer from fatigue during the second half and this is reflected in a drop in work rate (less distance covered and fewer sprints). A recent study using Pro-Zone technology to analyse time spent in different activities during 28 English Premier League matches found that in the last 15 min of a game, high-intensity running distance was ~20% less than in the first 15 min for outfield players in all positions, with similar deficits for high-intensity running with and without ball possession. There was also a noticeable decline (in high-intensity running immediately after the most intense 5-min period of the game with the greatest deficits (~40–50%) in attacking players and central defenders. It is common to see more goals scored in the later stages of games as players become tired and more mistakes are made. The development of fatigue during the game seems to be related to depletion of the muscle glycogen stores and it has been shown that players who started a match with a low glycogen content in their thigh muscles covered 25% less distance than the others (Table 6.1). Furthermore, players with a low initial muscle glycogen content covered 50% of the total distance walking and only 15% sprinting, compared with 27% walking and 24% sprinting for the players with normal to high muscle glycogen levels. Blood lactate concentration is consistently lower at the end of a game compared with values at half-time and this ties in with the observations that the greatest rate of decline in muscle glycogen occurs in the first half of the match. Players who start matches with low glycogen stores in their leg muscles are likely to be close to complete glycogen depletion by half-time

Muscle glycogen concentration (g/kg muscle ww)			Distance covered (km)			% Distance covered	
Before	Half-time	End	First half	Second half	Total	Walking	Sprinting
15	4	1	6.1	5.9	12.0	27	24
7	1	0	5.6	4.1	9.7	50	15

A group of Swedish professional soccer players was studied after playing a midweek game and preparing for another game on the Saturday. One part of the group was fed a high CHO diet for the few days between the games and the other group was allowed to eat their normal diet, which had a relatively low CHO content. Muscle biopsies were taken from the thigh muscles of the players before the Saturday game, at half-time and at the end of the match. Video analysis of the match was used to measure the distance covered by each of the players during the game. The fraction of the total distance covered at sprinting speed and walking speed was also determined, the remaining distance being covered at an intermediate speed. Note that the high CHO group had higher muscle glycogen stores at the start of the game, and at the end of the game still had some muscle glycogen left, whereas the other group had none. The total distance covered by the players in the first half of the game was not very different between the groups, but in the second half, when muscle glycogen concentration was lower, the players on the lower CHO diet were not able to run as far. The high CHO group covered more distance at sprinting speed (24% of a total distance of 12.0 km) and spent less time walking compared with the other group. Data from Saltin and Karlsson (1973).

Table 6.1 Muscle glycogen concentration and distance covered during the first and second half of a soccer match.

and these findings have important implications for the training and nutritional preparation of players. Until relatively recently these issues have largely been ignored.

Metabolic responses to intermittent high-intensity exercise

The metabolic responses to repeated sprints in soccer

A top-class soccer player has 150 to 250 brief intense actions during a game, indicating that the rates of PCr utilization and glycolysis are frequently high during a game, which is supported by findings of reduced muscle PCr levels and several-fold increases in blood and muscle lactate concentrations. Furthermore, the ADP degradation products—ammonia, hypoxanthine, and uric acid—are elevated in the blood during soccer matches. During intense periods of play muscle pH is lowered and muscle inosine monophosphate (IMP) elevated during a soccer game. Thus, the anaerobic energy systems are heavily taxed during periods of match-play. Glycogen in the working muscle seems to be the most important substrate for energy production during soccer matches. However, muscle triglycerides, blood free fatty acids, and glucose are also used as substrates for oxidative metabolism in the muscles. Blood glucose concentration, however, tends to be well maintained (or even above normal resting levels) during a game.

Plasma catecholamines are elevated and insulin lowered during a game. The plasma FFA levels increase progressively during a game, and are typically 2–3-fold higher than pre-exercise values at the end of a game, probably reflecting an increasing fat oxidation compensating for the lowering of muscle glycogen. Thus, elite soccer players have high aerobic requirements throughout a game and extensive anaerobic demands during periods of a match leading to major metabolic changes, which might contribute to the observed development of fatigue during and towards the end of a game.

The relative contribution of aerobic metabolism increases with increasing sprint number

At the onset of intense muscle contraction, a rapid and substantial hydrolysis of PCr and accumulation of lactate occur (see Chapter 3). When 30 s of high-intensity exercise is repeated with 4 min of resting recovery between successive bouts, the anaerobic contribution to the total amount of work done declines, while the aerobic contribution stays the same or increases slightly (see Figure 6.1). This means, of course, that the *relative* contribution of aerobic metabolism increases with increasing sprint number. In this situation it

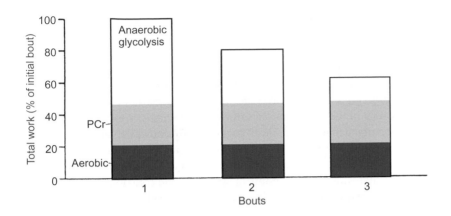

Figure 6.1

Estimated energy contribution during three bouts of maximal isokinetic cycling for 30 s at 100 rev/min with rest periods of 4 min between bouts. Darkened areas in bar graphs indicate the aerobic energy contribution. Data from Spriet *et al.* (1992).

appears that for the first few 30-s sprints at least the contribution of PCr hydrolysis can be maintained provided that the recovery time is sufficient to fully restore the PCr used in the previous bout. About 4 min is required for this. The decline in the anaerobic contribution in this situation is due to the fall in the contribution from anaerobic glycolysis. However, the relevance of this particular experimental protocol to the true sporting situation is questionable. Few sports involve maximal efforts lasting 30 s. In soccer, for example, sprints are sustained for only 2–3 s on average. Recently, the metabolic response to repeated sprints of this nature has been examined.

> Rates of muscle glycogen breakdown, PCr hydrolysis and lactate accumulation all decline substantially as the number of sprints performed increases

If all-out sprints of 6-s duration are repeated over several bouts, interspersed by rest periods of 30 s, the rates of muscle glycogen breakdown, PCr hydrolysis, and lactate accumulation all decline substantially as the number of sprints performed increases (data from an experiment illustrating this can be seen in Table 6.2 and Figure 6.2). Not surprisingly, this decreased energy turnover is accompanied by a decreased power output and a fall in the total amount of work done in successive sprints. The progressive fall in PCr utilization with successive sprints is probably related to the incomplete resynthesis of PCr between exercise bouts. If recovery is insufficient to enable complete PCr resynthesis to occur, anaerobic ATP resynthesis from PCr hydrolysis is limited during a subsequent bout of exercise. Other factors that are likely to have an impact on performance during multiple-sprint activities include the ability to re-establish skeletal muscle interstitial K^+ concentration and the intracellular concentrations of glycolytic intermediates, inorganic phosphate (P_i), and H^+ as these affect electrochemical–mechanical coupling. A gradual depletion of muscle glycogen stores is also likely to affect performance when intermittent high-intensity exercise is performed over a prolonged period. Many team games last 80–90 min or longer (with 'injury time' and 30 min extra time in some competitions). American football matches and professional singles tennis matches in Grand Slam events have been known to last over 4 h. During such prolonged periods of intermittent exercise, the general metabolic response is not very different from that observed for prolonged continuous submaximal exercise at a similar average work rate. Because of the higher rate of glycogenolysis in the active muscles during high-intensity exercise, it has been generally assumed that the muscle glycogen store becomes depleted more quickly during intermittent exercise compared with steady-state exercise at the same average power output. However, recent research has questioned this assumption as it seems that with repeated bouts of sprinting, the contribution of anaerobic glycolysis to energy turnover declines substantially with each successive bout.

Plasma glucose concentration seems to be well maintained during intermittent high-intensity exercise

Metabolite	First 6-s sprint		Tenth 6-s sprint	
	Pre-exercise	Post-exercise	Pre-exercise	Post-exercise
Glycogen	79	68	55	50
ATP	6.0	5.2	4.1	4.1
ADP	0.75	0.87	0.67	0.80
PCr	19.1	8.2	9.4	3.1
Cr	10.9	21.8	20.5	26.8
Glucose	0.4	0.6	2.0	2.1
Glucose 6-P	0.2	2.9	1.4	1.5
Fructose 6-P	0.05	0.60	0.45	0.35
Fructose 1,6-DP	0.05	0.40	0.10	0.13
TP	0.05	0.08	0.08	0.08
Pyruvate	0.15	0.50	0.40	0.45
Lactate	0.2	7.2	29.1	28.1

Values are means of eight subjects given in mmol glucosyl units/kgww for glycogen and mmol/kgww for all other metabolites. TP, triose phosphates. Data from Gaitanos *et al.* (1993).

Table 6.2 Muscle (vastus lateralis) metabolites before and after the first and tenth 6-s sprint on a cycle ergometer. 30 s resting recovery was allowed between each sprint.

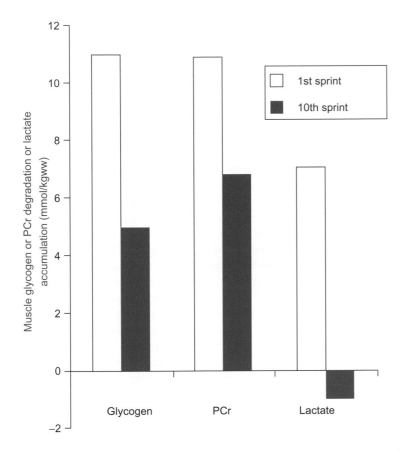

Figure 6.2

Rates of PCr degradation, glycogen degradation, and lactate accumulation in muscle during repeated sprints. Values were calculated from metabolite changes measured in biopsy samples obtained before and after the first and tenth 6-s sprint on a cycle ergometer. 30 s of resting recovery was allowed between each sprint. Data from Gaitanos *et al.* (1993).

During prolonged continuous exercise the blood glucose concentration tends to fall, but this does not seem to be the case for intermittent high-intensity exercise performed for 75–90 min. In fact, blood glucose can rise by 1–3 mmol/l during this type of activity. The blood glucose concentration reflects the balance between the rate of hepatic glucose release and the rate of peripheral uptake of glucose. Glycogenolysis in the liver is under hormonal control and is stimulated by glucagon and catecholamines, whereas insulin has the opposite effect. Plasma adrenaline concentration can reach fairly high levels during intermittent high-intensity exercise. During prolonged continuous steady-state exercise, the plasma insulin concentration falls, but in multiple-sprint activities marked increases in the plasma insulin concentration have been observed. The elevated insulin level in intermittent high-intensity exercise is thought to be secondary to the increased blood glucose concentration. Insulin inhibits lipolysis, and FFA re-esterification is promoted by high circulating levels of lactate (as noted earlier, blood lactate concentration is typically 4–6 mmol/l at the end of a game of soccer). These effects are likely to limit the mobilization and utilization of lipid fuels during multiple-sprint sports. Even so, the plasma concentrations of free fatty acids and ketone bodies (e.g. β-hydroxybutyrate) may double during an hour or more of intermittent high-intensity exercise and the circulating concentration of these substrates continues to increase for the first few hours of recovery if the individual remains in a fasted state.

> With repeated sprints the power output is increasingly supported by energy derived from PCr hydrolysis and aerobic metabolism, with a negligible contribution from anaerobic glycolysis

Studies on changes in intramuscular metabolites during multiple-sprint activity have revealed some interesting findings regarding the utilization of PCr and glycogen during this type of exercise. Data from one such study are shown in Table 6.2. In this experiment a group of games players performed ten 6-s maximal (all-out effort) sprints on a cycle ergometer with 30 s resting recovery between successive sprints. Needle biopsy samples were taken from the vastus lateralis muscle before and immediately after the first sprint and again 10 s before and immediately after the final sprint. Biochemical analysis of the biopsy samples revealed that the energy required to sustain the high mean power output that was generated over the first 6-s sprint (870 W) was provided by an equal contribution from PCr degradation and anaerobic glycolysis. Note that the mean power output during the first sprint was 2–3 times greater than that required to elicit VO_{2max} (~300 W). Immediately after the first sprint PCr concentration had fallen by 57% and muscle lactate concentration had risen by 7 mmol/kgww. However, in the tenth sprint there was no change in muscle lactate concentration even though mean power output had only fallen by 27% compared with that generated in the first sprint. The fall in muscle glycogen concentration during the tenth sprint was less than half of that observed during the first sprint and this reduced rate of glycogenolysis occurred despite a high plasma adrenaline concentration (which would be expected to stimulate glycogenolysis) after the ninth sprint. The change in PCr concentration was less during the tenth sprint than in the first sprint, but this was probably because the PCr concentration just before the tenth sprint was only half of its value before the first sprint. Hence, it can be concluded that during the tenth sprint, power output was supported by energy that was mainly derived from PCr hydrolysis and aerobic metabolism, with a negligible contribution from anaerobic glycolysis. The elevated intramuscular concentrations of free glucose and glucose 6-phosphate probably limit the contribution that blood glucose can make to energy turnover in multiple-sprint sports.

Resynthesis of phosphocreatine

> The resynthesis of PCr after high-intensity exercise follows an exponential curve and the time taken to resynthesize 50% of the resting store is about 30 s

The creatine kinase reaction is an equilibrium reaction (as is the adenylate kinase reaction) and is therefore reversible. This reaction occurs following exercise when the energy charge of the cell is increased and sufficient free energy is available to rephosphorylate Cr:

$$ATP + Cr \rightarrow ADP + PCr + H^+.$$

In general, the resynthesis of PCr after complete degradation of the muscle PCr store follows an exponential curve and the half-time for resynthesis (the time to resynthesize 50% of the resting store) is often quoted as 30 s (as illustrated in Figure 6.3) such that it takes about 4 min to restore PCr concentration to within 10% of pre-exercise values. In reality, however, there seems to be an enormous variation in the time course of resynthesis depending on the type of exercise performed and the duration and number of exercise bouts completed. Factors known to influence the rate of PCr resynthesis during recovery from exercise are the cellular concentrations of ATP, ADP, and Cr, which is not surprising given the equilibrium nature of the creatine kinase reaction. In addition, the hydrogen ion (H^+) is known to be a potent inhibitor of creatine kinase. In practice, therefore, a low muscle pH, a low oxygen tension, and/or a reduction in muscle blood flow will severely impair PCr resynthesis following exercise. Failure to fully restore PCr levels during recovery periods in intermittent exercise means that less PCr is available for ATP resynthesis at the start of the next bout. Several studies have demonstrated a relationship between the extent of PCr resynthesis and subsequent exercise performance.

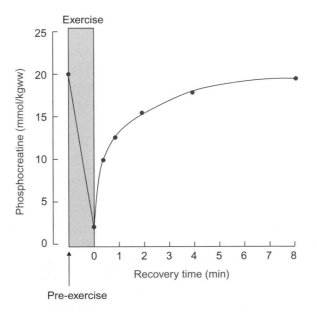

Figure 6.3

PCr resynthesis following high-intensity exercise.

Why does the rate of lactate production fall with repeated sprints?

Net lactate formation also occurs when the rate of pyruvate production by glycolysis exceeds the rate of pyruvate uptake into the mitochondria

The mechanism behind the fall in lactate production with repeated sprints is less clear. Lactate dehydrogenase is the enzyme that catalyses the following reversible reaction:

$$\text{Pyruvate} + \text{NADH} + \text{H}^+ \leftrightarrow \text{Lactate} + \text{NAD}^+.$$

The formation of lactate results from increasing pyruvate and NADH accumulation in the cytosol. When the supply of oxygen to tissues is inadequate, all tissues produce lactate by anaerobic glycolysis. Net lactate formation also occurs when the rate of pyruvate production by glycolysis exceeds the rate of pyruvate uptake (and NADH via the malate–aspartate shuttle) into the mitochondria, as well as the rate of conversion of pyruvate to alanine. Transamination of pyruvate to alanine, however, appears to be relatively insignificant during high-intensity exercise. Hence, it appears that the lower lactate production is most probably due to a reduced glycolytic rate when intense exercise is repeated after a short (a few minutes or less) recovery period. What is the cause of this? There are many possibilities (see Figure 6.4). As mentioned earlier it should be noted that the amount of work done in each repeated sprint falls as the number of sprints increases. In the study illustrated in Table 6.2 and Figure 6.2, the mean power output in the tenth 6-s sprint was 635 W compared with 870 W in the first, a fall of 27%. Thus, even in the tenth sprint, a power output equivalent to about 200% VO_{2max} was attained, although with this lower absolute power output, a higher proportion of the energy could be derived from aerobic metabolism.

End-product accumulation cannot completely explain the decline in lactate production seen during repeated bouts of contraction

It has been suggested that the fall in the rate of lactate production during repeated bouts of exercise occurs because of feedback inhibition of glycolysis caused by the accumulation of end-products such as H^+, P_i, and lactate ions. In addition to being a substrate for phosphofructokinase (PFK), ATP binds to a low-affinity, regulatory allosteric site of the enzyme. PFK is the rate-limiting enzyme of glycolysis, and the allosteric binding of ATP causes inhibition of the enzyme, thus slowing the rate of glycolysis. Most modulators of PFK activity appear to function by changing the binding of ATP to the allosteric site of PFK. However, several positive modulators of glycolysis (e.g. NH_4^+, ADP, and AMP) accumulate

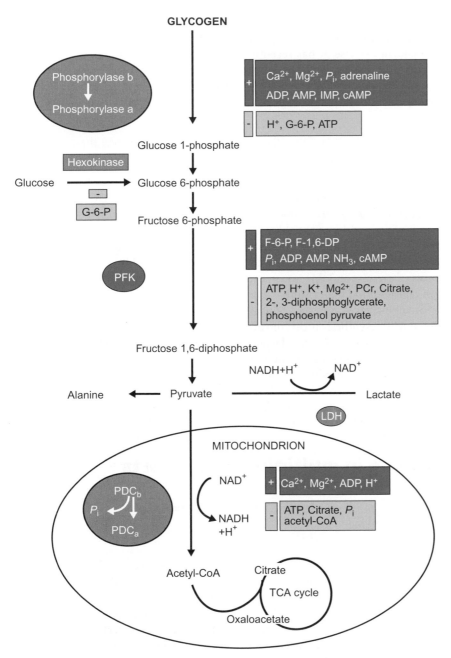

Figure 6.4

The major sites of regulation of the metabolic pathways related to carbohydrate utilization and lactate formation in skeletal muscle. Positive modulators are denoted with a plus (+), negative modulators with a minus (−).

in active skeletal muscle during high-intensity exercise, and it therefore seems that end-product accumulation cannot completely explain the decline in lactate production seen during repeated bouts of contraction. Furthermore, if the recovery interval between exercise bouts is extended, such that muscle metabolite concentrations are substantially diminished (i.e. they return close to values found in fresh resting muscle), there is still a reduction in lactate accumulation during subsequent exercise. It has recently been suggested that a progressive increase in flux through the pyruvate dehydrogenase complex (PDC)—which catalyses the conversion of pyruvate to acetyl-CoA—over the course of several bouts of exercise may be responsible for this observed decline in lactate accumulation.

A progressive increase in carbohydrate oxidation occurs over the course of several bouts of intense exercise

Calculations of the contribution of pyruvate flux through PDC to total ATP production during three bouts of 30 s of maximal exercise each separated by 4 min of recovery accounted for 29, 33, and 63% of total energy production in the first, second, and third bouts, respectively. This, of course, implies that a progressive increase in carbohydrate oxidation occurs over the course of several bouts of exercise and this is supported by the observed gradual increase in oxygen consumption under similar experimental conditions.

Small increases in oxygen consumption translate into large amounts of additional ATP resynthesis if carbohydrate is oxidized instead of being metabolized, without the use of oxygen, to lactate. Increasing the aerobic contribution to ATP resynthesis will also reduce the rate at which muscle glycogen is used up. The decline in lactate production during repeated bouts of maximal exercise also highlights an important point discussed in Chapter 4: attributing lactate production solely to an inadequate oxygen supply (i.e. muscle hypoxia) cannot explain the responses observed during repeated bouts of exercise.

Fatigue in multiple-sprint sports

Depletion of PCr specifically in Type-II muscle fibres may be primarily responsible for fatigue

In experiments where repeated bouts of maximal exercise have been performed with short recovery periods between exercise bouts, a significant relationship exists between the extent of PCr resynthesis between exercise bouts and subsequent exercise performance. Indeed, it has recently been demonstrated that when two bouts of 30-s maximal, isokinetic cycling exercise were performed and were separated by 4 min of recovery, the extent of PCr resynthesis during recovery was positively correlated with work output during the second bout of exercise. Furthermore, in agreement with the suggestion that the depletion of PCr specifically in Type-II muscle fibres may be primarily responsible for fatigue, it was demonstrated that the rate of PCr hydrolysis during the first bout of exercise was 35% greater in Type-II fibres than in Type-I fibres. However, during the second bout of exercise the rate of PCr hydrolysis declined by 33% in Type-II fibres (see Figure 6.5), which was attributable to incomplete resynthesis of PCr in this fibre type during recovery. Conversely, PCr resynthesis was almost complete in Type-I fibres during recovery from the first exercise bout and utilization was unchanged in this fibre type during the second exercise bout (Figure 6.5).

> Muscle glycogen availability is not usually considered to be responsible for fatigue during short-term high-intensity exercise

Anaerobic glycolysis is clearly an important source of ATP resynthesis during repeated bouts of high-intensity exercise, particularly in the first few bouts. Anaerobic glycolysis produces only three molecules of ATP for each molecule of glucose 6-phosphate derived from muscle glycogen compared with 38 molecules of ATP when the glucose molecule is completely oxidized to carbon dioxide and water. Even so, glycogen availability *per se* is not usually considered to be responsible for fatigue during short-term high-intensity exercise, provided that the pre-exercise glycogen store is not less than 25 mmol/kgww. However, some scientists have suggested that the critical level of muscle glycogen concentration below which impairment of anaerobic ATP resynthesis occurs is somewhat higher than this, at about 45 mmol/kgww. It is possible that glycogen availability will limit performance during repeated bouts of high-intensity exercise, if this type of activity is performed for a prolonged period, although this will depend to a large degree on the extent of the decline in the rate of glycogenolysis and

Figure 6.5

Phosphocreatine (PCr) utilization in Type-I and Type-II muscle fibres during two 30-s bouts of maximal isokinetic cycling. Each bout of exercise was separated by 4 min of resting recovery. Data from Casey *et al.* (1996a).

lactate production that occurs under these conditions. As described earlier, the initial muscle glycogen level of soccer players influences their performance (particularly in the second half of a game). A similar message comes from a study of ice-hockey players who raised their pre-exercise muscle glycogen content by 12% after dietary carbohydrate loading before competition. The group of players who glycogen-loaded covered greater distances during the game, and at faster average speeds than the control group.

Furthermore, in recent years, several studies have documented beneficial effects of ingesting carbohydrate solutions on soccer and tennis performance. It has been observed that during a soccer game most goals are scored towards the end of a game. This may occur because of a reduction in the work rate of the defenders or because of mental fatigue, leading to lapses in concentration and deterioration in skill. As blood glucose concentration does not decline during soccer-specific exercise protocols, it can be concluded that carbohydrate ingestion does not improve endurance performance and execution of skills in soccer by preventing the development of hypoglycaemia.

 Leaders in the field

Frederick Gowland Hopkins (1861–1947) studied chemistry at University College, London, before he graduated in medicine (at the age of 34) and taught physiology and toxicology for four years at Guy's Hospital. He moved to Cambridge to develop the chemical aspects of physiology. Biochemistry was not, at that time, recognized as a separate branch of science when Hopkins accepted the appointment. In 1902 he was given a readership in biochemistry, and in 1914 he was elected to the Chair of Biochemistry at Cambridge University. During all this time he had to be content with, at first, one small room in the Department of Physiology, and later with accommodation in the Balfour Laboratory; but in 1925 he was able to move his department into the new Sir William Dunn Institute of Biochemistry that had been built to accommodate it. Among his outstanding contributions to science was his discovery of a method for isolating tryptophan and for identifying its structure. Later, Hopkins worked with Walter Fletcher on the metabolic changes occurring in muscular contractions and rigor mortis. Hopkins supplied exact methods of analysis, and devised a new colorimetric reaction for lactic acid, and the pioneer work then done laid the foundations for the work of the Nobel Laureates, AV Hill and Otto Meyerhof, and also for that of many other later workers. In 1921 he was the first to isolate a substance that he named glutathione, and later he proved it to be the tripeptide of glutamic acid, glycine, and cysteine. He also discovered xanthine oxidase, a specific enzyme widely distributed in tissues and milk, which catalyses the oxidation of the purine bases xanthine and hypoxanthine to uric acid. Hopkins thus returned to the uric acids of his earliest work, a method of determining uric acid in urine, which he published in 1891.

In match play fatigue or reduced performance occurs at different stages of a game and may have different causal mechanisms

According to time–motion analyses and performance measures during match-play, fatigue or reduced performance seems to occur at three different stages in the game: (1) after short-term intense periods in both halves; (2) in the initial phase of the second half; and (3) towards the end of the game. Temporary fatigue after periods of intense exercise in the game does not appear to be linked directly to muscle glycogen concentration, lactate accumulation, acidity, or the break-down of PCr. Instead, it may be related to disturbances in muscle ion homeostasis and an impaired excitation of the sarcolemma. Soccer players' ability to perform maximally is impaired in the initial phase of the second half, which may be due to lower muscle temperatures compared with the end of the first half. Thus, when players perform low-intensity activities in the interval between the two halves, both muscle temperature and performance are better preserved. Muscle glycogen is typically reduced by 40% to 90% during a game and is probably the most important substrate for energy production. Even when whole muscle gly-cogen may appear adequate, fatigue towards the end of a game might be related to depletion of glycogen in some individual muscle fibres that have been active during the game. In a recent study it was found that whole-muscle glycogen decreased by about 43% during a soccer game, but that almost half of the muscle fibres were completely or almost empty of glycogen after the game (Figure 6.6).

In a hot and humid environment, dehydration and hyperthermia may result in reduced cerebral function that may also contribute to the deterioration in performance. This may be related to changes in synthesis and release of key neurotransmitters but may also be related to changes in the permeability of the

Figure 6.6

Mean values from 10 players for relative glycogen content in Type-I, IIa, and IIX fibres and in all fibres before and immediately after a soccer match. Data from Krustrup *et al.* (2006).

blood–brain barrier. In conclusion, fatigue or impaired performance in soccer (and probably most other sports involving multiple sprints) occurs during various phases in a game, and different physiological mechanisms appear to operate in different periods of a game.

Inorganic phosphate (HPO_4^{2-}) accumulation may be an important factor in human muscle fatigue

One of the consequences of rapid PCr hydrolysis during high-intensity exercise is the accumulation of P_i, which has been shown to inhibit muscle contraction coupling directly. However, the simultaneous depletion of PCr and P_i accumulation makes it difficult to separate the effect of PCr depletion from P_i accumulation *in vivo*. This problem is further confounded by the parallel increases in H^+ and lactate ions that occur during high-intensity exercise. All of these metabolites have been independently implicated with muscle fatigue. Recently, studies have been conducted on the metabolic basis of human muscle fatigue and recovery using ^{31}P-nuclear magnetic resonance spectroscopy and measurements of the force generated. One experimental model has used a 4-min period of maximum voluntary contraction (MVC) in two different muscles: adductor pollicis (the small muscle that is used to adduct the thumb, i.e. move it towards the hand) and tibialis anterior (a muscle at the front of the lower leg used in dorsiflexion of the ankle and inversion of the foot). For both muscles, during fatiguing exercise, a non-linear relationship between the force that could be achieved in an MVC and both PCr and total inorganic phosphate (P_i) was observed. By contrast, there was a roughly linear relationship between the decline in MVC force and the accumulation of both H^+ and the monovalent species of inorganic phosphate (HPO_4^{2-}). However, during recovery after exercise, MVC force rapidly returned to pre-exercise levels while the H^+ concentration recovered with a much slower time course. HPO_4^{2-} concentration, however, rapidly returned to pre-exercise values with a time course similar to MVC force recovery. In addition, the relationship between HPO_4^{2-} concentration and MVC force was similar during both fatigue and recovery. Thus, during fatigue as well as recovery, changes in MVC force correlate best with HPO_4^{2-}, suggesting that this metabolite is an important factor in human muscle fatigue. Preliminary evidence from *in vitro* studies of isolated animal muscle fibres suggests that HPO_4^{2-} directly inhibits cross-bridge formation.

Altered calcium transport kinetics may be implicated in fatigue

As described in Chapter 2, calcium release by the sarcoplasmic reticulum as a consequence of muscle depolarization is essential for the activation of muscle excitation–contraction coupling. It has been demonstrated that during fatiguing contractions there is a slowing of calcium transport and progressively smaller calcium transients that has been attributed to a reduction in calcium re-uptake by the sarcoplasmic reticulum and/or increased calcium binding in the cytoplasm.

Strong evidence that a disruption of calcium handling is responsible for fatigue comes from studies showing that stimulation of sarcoplasmic reticulum calcium release caused by the administration of caffeine to isolated muscle can improve muscle force production, even in the face of a low muscle pH. Thus, there may be a long-term effect of prior high-intensity exercise on calcium ion handling by the sarcoplasmic reticulum.

Several studies have shown that reduced muscle pH can interfere with excitation–contraction coupling, and high muscle lactate concentrations have also been reported to inhibit calcium ion release from the sarcoplasmic reticulum. Therefore, the removal of lactate and H^+ from skeletal muscle is likely to be of importance for the ability to maintain power output during repeated bouts of high-intensity exercise. Export of lactate across the sarcolemmal membrane is mediated by a lactate/H^+ co-transporter. A number of isoforms of these mono-carboxylic acid transporter (MCT) proteins have been identified in both animal and human muscle. High-intensity exercise training enhances the lactate/H^+ transport capacity of human skeletal muscle as well as increasing the levels of MCT1 and MCT3. Light to moderate muscular activity increases the rate at which lactate is eliminated from the muscle and circulation during recovery from high-intensity exercise. Lactate is taken up from the blood mainly by the liver, heart, and Type-I skeletal muscle fibres. Most of it is converted to pyruvate and oxidized by these tissues.

It has been demonstrated that when muscle fatigue is induced using electrical stimulation, recovery takes significantly longer to occur when low-frequency electrical stimulation (e.g. 20 Hz) is used rather than high-frequency stimulation (e.g. 100 Hz). This delayed recovery has been attributed to disturbances in excitation–contraction coupling and, as muscle activation by motor neurons *in vivo* is in the low-frequency range, this response could offer some insight into the mechanism of fatigue development. Prior high-intensity exercise may cause a more rapid reduction in neural motor drive in subsequent high-intensity exercise bouts.

Nutritional strategies for team sports athletes

Energy and carbohydrate needs

> Carbohydrate is an important energy source in both training and competition

Team sports athletes are often less aware of the need for attention to diet to support consistent training and to ensure optimum performance in match play. Many team sports are characterized by an intensive competition schedule that may involve two or three competitions per week (even more for some basketball or baseball players) for a large part of the year, with an off-season in which little or no exercise is undertaken. Pre-season training is often intensive, with the aim of reducing body weight, and especially body fat content, as well as improving fitness. The games player requires strength and power as well as stamina, so the training programme is especially challenging. There may also be a need, especially with younger players, to increase muscle mass: as this normally requires a positive energy balance, it becomes especially difficult when simultaneous reduction of fat mass is required.

The high demands on carbohydrate as an energy source in both training and competition make it important that players ensure an adequate dietary carbohydrate intake, but this must be achieved within the available energy budget. This means moderating the intake of fat and protein to limit total energy intake. It also means paying special attention to the timing of carbohydrate intake: eating soon after training or competition will promote rapid glycogen synthesis. Choosing foods with a high glycaemic index—those that cause a rapid elevation of blood glucose levels and stimulate a marked insulin response—at this time will also help to ensure rapid storage of the ingested carbohydrate in the form of muscle glycogen.

Consumption of a high carbohydrate diet in the days prior to competition may benefit competitors in games such as rugby, soccer, or hockey, where the multiple short sprints that are performed rely on anaerobic metabolism. It appears not to be usual for these players to pay attention to this aspect of their diet, but, as explained earlier in this chapter, a Swedish study in the early 1970s showed that players starting a soccer game with low muscle glycogen content did less running, and much less running at high speed, than those players who began the game with a normal muscle glycogen content (details can be found in Table 6.1). It is common for players to have one game in midweek as well as one at the weekend, and it is likely that full restoration of the muscle glycogen content will not occur between games unless a conscious effort is made to achieve a high carbohydrate intake.

Players may benefit from ingestion of some carbohydrate during a match: this may be especially true if there has been limited recovery time since the last match resulting in muscle and liver glycogen stores that are less than optimal. The availability of ingested carbohydrate may be limited, however, as the short bursts of high intensity effort that occur in sports such as soccer seem to be sufficient to cause some delay in the rate of gastric emptying of ingested drinks. If the carbohydrate remains in the stomach, it will not be available as a fuel for the working muscles. There is now some evidence, though, that carbohydrate receptors are present in the mouth and that stimulation of these receptors can act—presumably on the brain—to lessen the sensation of fatigue and to improve performance.

Few of the supplements promoted to enhance loss of body fat are effective, and those that are effective are mostly based on stimulants such as ephedrine. Apart from the health risks associated with the use of these products, they are prohibited by the doping regulations and players who test positive are liable to suspension from competition.

Creatine supplements and performance of multiple sprints

Creatine supplementation allows more work to be performed during repeated bouts of maximal exercise

A growing body of evidence is becoming available to indicate that dietary creatine (Cr) intake may be a necessary requirement for individuals wishing to optimize their ability to perform repeated bouts of high-intensity exercise. Creatine, or methyl guanidine-acetic acid, is a naturally occurring compound and the total body creatine pool in man amounts to approximately 120 g, of which 95% is found in muscle. In human skeletal muscle, Cr is present at a concentration of about 30 mmol/kgww, of which approximately 60% is in the form of PCr in resting muscle. In normal healthy individuals, muscle Cr degrades irreversibly to creatinine at a rate of approximately 2 g/day but is continually replenished by endogenous Cr synthesis from amino acids in the liver and/or dietary Cr intake, mostly from meat.

Five days of Cr ingestion (20 g/day) has been shown to significantly increase the amount of work that can be performed by healthy normal volunteers during repeated bouts of maximal exercise. This conclusion is based on the results from laboratory studies involving repeated bouts of maximal dynamic and isokinetic cycling exercise (e.g. three bouts of maximal cycling exercise each separated by 4 min resting recovery) and from controlled 'field' experiments undertaken by athletes (e.g. 4 × 300-m running interspersed with 4 min recovery). The consistent finding from these studies is that Cr ingestion significantly increased exercise performance by sustaining force or work output during successive bouts of high-intensity exercise, such that total work production during exercise increased by 5–7%. The mechanisms by which performance is improved are unclear. However, the improvement is unlikely to be solely a consequence of an increase in pre-exercise PCr availability, as the magnitude of this increase (~2 mmol/kgww or ~10%) seems to be insufficient to produce the improvements reported.

Improved performance in repeated sprints may be due to a stimulatory effect of Cr ingestion on PCr resynthesis during recovery periods

The enhancement of performance may also be related to a stimulatory effect of Cr ingestion on PCr resynthesis during exercise and recovery. Considering that PCr availability is thought to limit exercise performance during maximal exercise, all of these effects would increase muscle contractile capability by maintaining anaerobic ATP turnover during exercise. This suggestion is supported by findings that have shown that Cr supplementation can reduce plasma ammonia and hypoxanthine accumulation (which are indicative of adenine nucleotide degradation) during maximal exercise, while at the same time increasing work output.

Figure 6.7

(a) Individual values for muscle total creatine (TCr) concentration before and after 5 days of creatine (Cr) ingestion (4 × 5 g/day). Subjects have been numbered 1–8, based on their initial muscle TCr content. (b) Individual increases in muscle TCr content after Cr ingestion for the same subjects depicted in (a), plotted against the change in phosphocreatine (PCr) resynthesis during recovery after Cr ingestion. Data from Greenhaff *et al.* (1994).

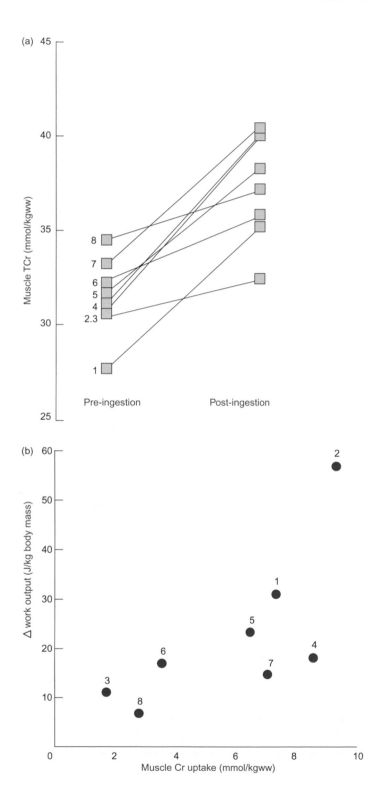

Figure 6.8

(a) Muscle total creatine (TCr) concentration in individual subjects pre- and post-creatine (Cr) supplementation for 5 days (4 × 5 g/day). Subjects have been numbered 1–8 based on their initial muscle TCr concentration. (b) Individual increases in muscle TCr after Cr ingestion for the same subjects depicted in (a), plotted against the increase in work production over two bouts of maximal isokinetic cycling exercise. Data from Casey et al. (1996b).

Considering recovery from high-intensity exercise, it is known that individuals who demonstrate more than a 25% increase in muscle total Cr concentration as a result of Cr supplementation can expect to experience an accelerated rate of PCr resynthesis during recovery (see Figure 6.7). It is likely that this occurs as a result of Cr ingestion maintaining the muscle free Cr concentration higher than the K_m of mitochondrial creatine kinase (CK) for Cr (15 mmol/kgww) throughout recovery, thereby sustaining a high flux through the CK reaction in favour of PCr resynthesis.

Significant points often overlooked when attempting to use Cr supplementation to improve maximal exercise performance are, first, that not everyone responds to supplementation; muscle Cr uptake appears to be relatively low in 30% of individuals. Secondly, the most startling effects of Cr supplementation on exercise performance are usually observed in individuals who experience more than a 25% increase in muscle total Cr during supplementation (see Figure 6.8). With these points in mind, it is important to note that recent work has demonstrated that Cr ingested in combination with carbohydrate can increase Cr retention in all subjects by more than 25%. This phenomenon appears to be due to an insulin-stimulated increase in muscle Cr uptake.

KEY POINTS

1. Games such as soccer, rugby, and hockey involve repeated bouts of high-intensity exercise interspersed with short periods of relatively lighter exercise. The average exercise intensity during play may be as high as 75% VO_{2max}.

2. Maintaining performance for the full duration of matches is obviously important to success and depends to a large degree on the ability of muscle to recover from the last exercise bout. The extent of recovery is affected by the intensity and duration of the last bout, the biochemical and physiological characteristics of the individual, and the time available for recovery before another bout of exercise has to be performed.

3. At the onset of intense muscle contraction, a rapid and substantial hydrolysis of PCr and accumulation of lactate occur. When 30 s of high-intensity exercise is repeated with 4 min of resting recovery between successive bouts, the anaerobic contribution to the total amount of work done declines, while the aerobic contribution stays the same or increases slightly. This means that the relative contribution of aerobic metabolism increases with increasing sprint number.

4. When repeated bouts of maximal exercise are performed, the rates of muscle PCr hydrolysis and lactate accumulation decline. In the case of PCr, this response is thought to occur because of incomplete PCr resynthesis occurring during recovery between exercise bouts. However, the mechanism(s) responsible for the fall in lactate accumulation is unclear. It has been suggested that a progressive increase in flux through the pyruvate dehydrogenase complex (PDC)—which catalyses the conversion of pyruvate to acetyl-CoA—and a progressive increase in carbohydrate oxidation occurs over the course of several bouts of exercise are responsible for this observed decline in lactate accumulation.

5. Other factors that are likely to have an impact on performance during multiple-sprint activities include the ability to re-establish skeletal muscle interstitial K^+ concentration and the intracellular concentrations of glycolytic intermediates, inorganic phosphate (Pi), and H^+ as these affect electrochemical–mechanical coupling. A gradual depletion of muscle glycogen stores is also likely to affect performance when intermittent high-intensity exercise is performed over a prolonged period.

6. A growing body of evidence is becoming available to indicate that dietary creatine (Cr) intake may be a necessary requirement for individuals wishing to optimize their ability to perform repeated bouts of high-intensity exercise.

7. Oral creatine supplementation of 20 g/day for 5–6 days increases the muscle total creatine content in men by about 20% (30–40% of the increase is PCr). A subsequent daily dose of 2 g is enough for maintenance of this increased concentration.

8. Ingesting creatine in combination with carbohydrate accelerates creatine loading. Creatine loading allows an increase in the amount of work performed during single and repeated bouts of short-lasting high-intensity exercise. The mechanism behind this effect is probably a decreased lactate production (glycolysis) and higher ATP turnover due to the higher initial PCr concentration and a faster PCr resynthesis rate during recovery between bouts.

9. The few studies that have investigated the combined effects of long-term creatine ingestion and strength or high-intensity running training show impressive improvements in performance, greater than those seen after short-term supplementation. This may be explained both by the possibility that the higher creatine content enables the athletes to do more repetitions during training at a higher speed or workload and by the possibility that creatine, via a mechanism involving muscle fibre volume swelling, exerts a direct anabolic effect leading to improved adaptation and thus improved sprint performance, maximal strength, and high-intensity endurance performance.

10. Besides weight gain (typically 1–2 kg) due to osmotic retention of water, creatine does not seem to have major side-effects. However, the long-term effects of creatine ingestion are unknown.

Selected further reading

Bangsbo J *et al.* (2007). Metabolic response and fatigue in soccer. *International Journal of Sports Physiology and Performance* 2(2): 111–127.

Birch R *et al.* (1994). The influence of dietary creatine supplementation on performance during repeated bouts of maximal isokinetic cycling in man. *European Journal of Applied Physiology* 69: 268–270.

Bogdanis GC *et al.* (1996). Contribution of phosphocreatine and aerobic metabolism to energy supply during repeated sprint exercise. *Journal of Applied Physiology* 80: 876–884.

Bradley PS *et al.* (2009). High-intensity running in English FA Premier League soccer matches. *Journal of Sports Sciences* 27: 159–168.

Casey A *et al.* (1996a). The metabolic response of type I and II muscle fibres during repeated bouts of maximal exercise in man. *American Journal of Physiology* 271: E38–E43.

Casey A *et al.* (1996b). Effect of creatine supplementation on muscle metabolism and exercise performance. *American Journal of Physiology* 271: E31–E37.

Casey A *et al.* (1996c). The effect of glycogen availability on power output and the metabolic response to repeated bouts of maximal isokinetic exercise in man. *European Journal of Applied Physiology* 72: 249–255.

Ekblom B (1986). Applied physiology of soccer. *Sports Medicine* 3: 50–60.

Gaitanos GC *et al.* (1993). Human muscle metabolism during intermittent maximal exercise. *Journal of Applied Physiology* 75: 712–719.

Greenhaff PL (2000). Creatine. In: *Nutrition in sport* (ed. Maughan RJ). Oxford: Blackwell, pp. 379–392.

Greenhaff PL *et al.* (1993). Influence of oral creatine supplementation on muscle torque during repeated bouts of maximal voluntary exercise in man. *Clinical Science* 84: 565–571.

Greenhaff PL *et al.* (1994). The effect of oral creatine supplementation on skeletal muscle phosphocreatine resynthesis. *American Journal of Physiology* 266: E725–E730.

Hargreaves M (ed.) (1995). *Exercise metabolism*. Champaign, IL: Human Kinetics.

Hultman E and Greenhaff PL (1996). The metabolic responses of type I and II muscle fibres during repeated bouts of maximal exercise in man. *American Journal of Physiology* 271: E38–E43.

Jacobs I *et al.* (1982). Muscle glycogen and diet in elite soccer players. *European Journal of Applied Physiology* 48: 297–302.

Kirkendall DT (1993). Effects of nutrition on performance in soccer. *Medicine and Science in Sports and Exercise* 25: 1370–1374.

Krustrup P *et al.* (2006). Muscle and blood metabolites during a soccer game: Implications for sprint performance. *Medicine and Science in Sports and Exercise* 38: 1165–1174.

Medbo JI and Tabata I (1989). Relative importance of aerobic and anaerobic energy release during short-lasting exhausting bicycle exercise. *Journal of Applied Physiology* 67: 1881–1886.

Mohr M *et al.* (2005). Fatigue in soccer: a brief review. *Journal of Sports Sciences* 23: 593–599.

Nicholas CW *et al.* (1995). Influence of ingesting a carbohydrate-electrolyte solution on endurance capacity during intermittent, high intensity shuttle running. *Journal of Sports Sciences* 13: 283–290

Reilly T (1996). Motion analysis and physiological demands. In: *Science and soccer* (ed. T Reilly). London: E & FN Spon, pp. 65–81.

Saltin B and Karlsson J (1973). Die Ernahrung des Sportlers. In: *Zentrale Themen de Sportmedizin* (ed. Hollman W). Berlin: Springer, pp. 245–260.

Spriet LL *et al.* (1989). Muscle glycogenolysis and H^+ concentration during maximal intermittent cycling. *Journal of Applied Physiology* 66: 8–13.

Walsh NP *et al.* (1998). The effects of high-intensity intermittent exercise on the plasma concentrations of glutamine and organic acids. *European Journal of Applied Physiology* 77: 434–438.

Williams MH *et al.* (1999). *Creatine: the power supplement*. Champaign, IL: Human Kinetics.

Sporting talent: the genetic basis of athletic capability

7

Learning objectives

After studying this chapter, you should be able to …

1. appreciate the different factors that can determine success in sport;
2. describe the structure and composition of the genetic material (DNA);
3. describe the processes of transcription and translation;
4. understand how different proteins can be derived from the same gene;
5. understand the principles of heredity;
6. discuss whether muscle fibre-type composition is determined by nature or nurture;
7. appreciate the genetic limitations to sporting ability and the capacity to adapt to a training stimulus.

Introduction: factors determining success in sport

Success in sport is determined by many factors, including motivation, appropriate training, nutrition, tactics and, perhaps most importantly, raw talent

A person's genetic make-up is called his or her genotype. The physical expression of the genotype as particular characteristics or traits (e.g. height, strength, hair colour, etc.) is called the person's phenotype. Success in sport is determined by many factors, including motivation, appropriate training, nutrition, and tactics. However, perhaps the most important factor is raw talent in terms of the body's phenotype; in other words the body's physical, physiological, and metabolic characteristics. These characteristics, which in terms of athletic capability might be taken to include muscle fibre-type composition, the size of the heart and lungs, body height and mass, etc., are all to a large extent determined by the genotype (or genetic endowment) of the individual.

An individual's maximum oxygen uptake is modifiable by training but is mostly determined by genetics

A good example to look at in the sporting context is an individual's maximum oxygen uptake (VO_{2max}), which is strongly related to endurance performance. The VO_{2max} is determined to a large extent by the pumping capacity of the heart (i.e. the cardiac output), which in turn is related largely to the stroke volume and hence the size of the heart (maximal heart rates are similar in athletes and non-athletes). The VO_{2max} of 18–35-year-old elite endurance athletes is usually at least 70 ml/kg/min and perhaps as high as 90 ml/kg/min (in men), whereas that of similarly aged sedentary individuals and sports men or women whose sports do not require high endurance capacity may be only 35–50 ml/kg/min. Thus, it is possible that two individuals of the same age and gender can have VO_{2max} values that differ by more than 100%. While appropriate exercise training can increase VO_{2max}, such improvements usually amount to less than 20% of the pre-training VO_{2max}. This example emphasizes the importance of the genetic component to success in athletic endeavour.

Of course, other factors such as nutrition and willingness to undergo the necessary training are important to allow the full development of the genetic potential and it is also obvious that most sporting skills decline with advancing age. Other aspects of sporting talent, such as hand–eye coordination—so important in racquet sports—can also be developed by a lifetime of practice, but even with a motor skill such as this there is also probably a significant genetic component in terms of visual acuity and tracking a moving object, as well as the ability to learn and execute complex movements.

Certain physical characteristics are essential for success in many sports at the elite level (for example, in the past 20 years, no male tennis player under 1.75 m in height has won a Grand Slam event and American National Football League linebackers weighing less than 90 kg are a rarity!). These physical characteristics are determined to a large degree by the genetic information that is carried by each and every one of us. Only monozygotic twins, individuals who develop from the same fertilized ovum (known as the zygote) as a result of the splitting of the cell mass at a very early stage of embryonic development, carry exactly the same genetic information. Non-identical (dizygotic) twins result from the fertilization of two different ova, and hence have a different genotype.

The nature of the genetic material

The information required for the synthesis of specific proteins is encoded in DNA

All the genetic information of every species is contained in its deoxyribonucleic acid (DNA) structure, and this determines the type and amount of protein synthesized in each cell of the organism. These proteins are in turn responsible for the synthesis of all other cellular components: the genetic material codes only for proteins and does this by defining their component amino acids. Proteins provide the structural basis of all tissues and organs, and it is largely the protein content of these tissues that gives them their recognizable shape. More importantly, perhaps, the proteins present in the different tissues confer on each tissue its metabolic capabilities. The presence or absence of a particular enzyme determines whether or not a tissue can carry out a particular function, and the activity (which depends on the amount of enzyme or an isoform) determines how fast that process can proceed. Proteins and amino acids also constitute, or act as precursors for, many of the body's hormones, regulatory peptides, and neurotransmitters as well as acting as the receptors for these signalling systems and fulfilling a variety of other functions.

All somatic cells of the body contain the same genetic material in their nuclei, but not all genes are expressed

Although all somatic cells of the body (that is all cells except the germ cells, sperm, and ova) contain the same genetic material in their nuclei, not all genes are expressed (i.e. available to be translated into protein). Thus, the structural and functional characteristics of different cell types are determined by selective gene expression. Although all cells in the human body express certain genes (e.g. those that code for the enzymes of glycolysis), only some cells express genes for other specific proteins (e.g. myosin, troponin, hormone receptors, or enzymes of a metabolic pathway specific to a particular tissue type) and other genes are repressed. This is what makes a liver cell different from a muscle cell or a nerve cell. Alteration of gene expression is one of the means by which the body develops and adapts. Certain hormones (particularly steroid hormones such as cortisol and testosterone) are known to be important in the regulation of gene expression.

Nucleic acids and control of protein synthesis

The parts of the DNA that code for specific proteins are called genes

The development of the cell is determined by the chromosomes that are present in its nucleus, and that contain the genetic information that defines the characteristics of the mature cell by regulating the synthesis of the many thousands of different proteins that give the cell its structural and functional characteristics. Chromosomes are a compact form of DNA complexed with a protein called chromatin and only appear just prior to cell division. At other times in the cell cycle DNA in the nucleus is in an uncoiled form and when freed from contact with chromatin can be used as a template for ribonucleic acid (RNA) and then protein synthesis. The parts of the DNA that code for specific proteins are called genes.

The genome is the genetic fingerprint of an organism

The genome is the entire DNA sequence of an organism and so essentially represents the genetic fingerprint of an individual. The term genome can also be applied to all the variant forms of genes that exist in a population or species, and this can be termed the species genome. The Human Genome Project is the largest international collaboration in biology to date and has resulted in the sequencing of three billion coding units in human DNA, though most genes and their functions have not yet been definitely identified. Human cells are thought to contain about 35 000 genes and there are sequence similarities between many genes of related functions in different species. The exact number of genes in the human body is still a little unclear but the estimate of around 35 000 emerged when the human genome sequence was 'cracked' as published in simultaneous papers in Nature (Human Genome Project) and Science (Celera Genomics). Prior to these publications, it was felt that there might be as many as 100 000 genes in the human body and this relatively small number came as a surprise to many scientists. These genes are organized in relatively small clusters throughout the entire DNA with extremely large tracks of non-coding DNA intervening between each gene. The genomes of many other organisms also contain more than just sets of genes. Some 90–95% of human DNA does not seem to code for anything and was until recently commonly referred to as 'junk DNA'. However it is now known that areas of 'non-coding DNA' can actually encode for modifying forms of RNA called micro-RNAs that can influence mRNA abundance and other changes in non-coding regions

(i.e. tri- and tetra-nucleotide expansions) can influence gene expression in surrounding regions. Furthermore, each single gene has its coding sequences interrupted by non-coding sequences called introns. The scattered pieces of coding
sequences, called exons, are usually shorter than the introns such that the coding
portion of a gene is often only a small fraction of the length of a gene.

Chromosomes, which are simply packages of DNA, are paired; therefore,
there are two copies of each gene and two sets of genetic instructions, one from
the maternal chromosome and the other from the paternal chromosome. The
matching genes are always in the same position, or locus, on a particular chromosome and are called alleles. Alleles may code for the same or alternative forms
of a trait. Genes coding for development (two arms, two legs, sexual maturation,
etc.) are always identical on maternal and paternal chromosomes, but in the
case of some particular features (e.g. eye colour, tongue rolling, etc.) one allele
may mask the other and is said to be dominant. The allele that is masked is said
to be recessive and its message is not expressed, but it remains present on the
chromosome and may be passed to the next generation in a gamete. If both genes
are identical, the person is said to be homologous for that trait; if the genes are
different, the person is said to be heterozygous for that trait. In this case the trait
expressed will be that of the dominant allele.

Techniques

Western blotting is a particular form of immunoblotting. These techniques rely
on the specificity of binding between the target molecule and a specific probe
that reacts with this molecule and allows detection of the molecule of interest
in a mixture of many other similar molecules. In Western blotting, the molecule
of interest is a protein and the probe is typically an antibody raised against that
particular protein. The method allows the detection of specific proteins in biological samples and is widely used in the analysis of blood and tissue extracts.
The technique is based on gel electrophoresis and allows separation based on
either the length of the polypeptide chain after denaturation of the native protein or on the 3D structure of the intact protein. Gel electrophoresis uses an
electric current applied to a gel matrix to cause migration of the proteins from
the point of application to the gel plate. The speed of migration of the proteins
in the sample is determined primarily by their size. After separation, the proteins
are allowed to react with specific antibodies, and the amount of protein is quantified by colorimetric, fluorometric, chemiluminescent, or radioactive detection. The name Western blot is derived from the term Southern blot, which is a
technique for detection of DNA. The Southern blot was named after Professor
Sir Edwin Southern, who is currently Professor of Biochemistry (Emeritus) at the
University of Oxford. Northern blotting is used for the detection of RNA, and
Eastern blotting for detection of post-translational modification of proteins.

Four different nucleotide bases are present in DNA

All human somatic cells contain 23 pairs of chromosomes and there are thousands of different proteins in each cell. Thus, there are many genes on each chromosome. The functional unit of DNA—a deoxyribonucleotide—consists of a pentose sugar molecule (deoxyribose), a phosphate group, and an organic base that is either a purine or a pyrimidine. The four different bases that are present in DNA are adenine (A), thymine (T), guanine (G), and cytosine (C). Adenine and thymine are purines, while guanine and cytosine are pyrimidines. The backbone of the molecule consists of two antiparallel chains of alternating deoxyribose and phosphate groups, and the DNA molecule is typically tens of millions of these units long: the structure of a short segment of a single chain containing each of the bases is shown in Figure 7.1. The order of the nucleotide bases in DNA

Figure 7.1

A short section of one of the strands of a DNA molecule, showing the backbone consisting of alternate deoxyribose and phosphate groups and the attachment of the four different bases.

determines the order of the amino acids in the protein that will be synthesized, and the process is switched on and off by control sequences.

DNA replication occurs during the process of cell division

The first step in cell division is the replication of the DNA in the cell that is about to divide into two daughter cells. DNA replication begins by first synthesizing a new strand of DNA on each of the pre-existing strands. The latter act as templates and so each new strand that is synthesized is a complementary strand to, not a copy of, the original strand. Various initiation factors and enzymes may initiate replication and DNA synthesis is usually primed by a short strand of RNA that is transcribed onto the DNA template (as described below). During this process, ribonucleotides in the RNA strand form base pairs with nucleotides in a DNA template strand and Pi is released. Termination of DNA synthesis occurs when the entire duplex is replicated.

Transcription

The process of transcription is the formation of a complementary strand of RNA based on the DNA template

The chemistry of the bases in DNA allows bonding to occur between pairs of bases, with strong bonds being formed only between adenine and thymine and between guanine and cytosine (as you can see in Figure 7.2), and this accounts for the two parallel strands that run effectively in opposite directions forming a double helix. The hydrogen bonds that are formed are extremely stable, and this accounts for the stability of the genetic information that these molecules contain, but they can be broken during the process of transcription—the formation of a complementary strand of nucleic acids based on the DNA template—by which the information they contain is transferred to the protein synthetic apparatus.

During the process of transcription, the hydrogen bonds joining the bases are broken and the enzyme RNA polymerase forms a sequence of ribonucleotides,

Permissible pairing:

Thymine   Adenine   Cytosine   Guanine

Figure 7.2

The specific bonding of thymine to adenine and cytosine to guanine means that the structure of one of the strands of DNA determines the composition of the other.

following the same base-pairing arrangement with the exception of the presence of uracil (U) in RNA rather than the thymine present in DNA. The sequence of bases in the original DNA molecule (or at least in one strand of it—the other strand is not used—thus determines the order of bases on the molecule of RNA, known as messenger RNA (mRNA), as shown in Figure 7.3. In other words the DNA serves as a template from which a complementary RNA molecule is transcribed. The entire length of the gene including introns as well as the exons is first transcribed into a long RNA molecule, the primary transcript. After capping at the 5′ end (the end that was synthesized first during transcription) with an atypical nucleotide (guanine with a methyl group attached) and the addition of a polyadenylate group at the 3′ end of the primary transcript, but before the RNA leaves the nucleus, all of the intron sequences are removed and the exons are joined together. The result is a much shorter RNA molecule that now contains an uninterrupted coding sequence. This step is called RNA splicing and after it has been completed, the RNA is a functional mRNA molecule that can now leave the nucleus and be translated into protein (Figure 7.4). The introns are removed from RNA by enzymes that recognize a few specific short nucleotide sequences that act as cues for its removal. These sequences are found at each end of the intron and are the same or very similar in all introns. Although the intron and exon type of gene arrangement seems at first sight rather wasteful, the primary transcripts of many genes can be spliced in various ways to produce different mRNAs, depending on the cell type or the stage of development as illustrated in Figure 7.5. This process, known as alternative RNA splicing, allows several different proteins to be produced from the same gene.

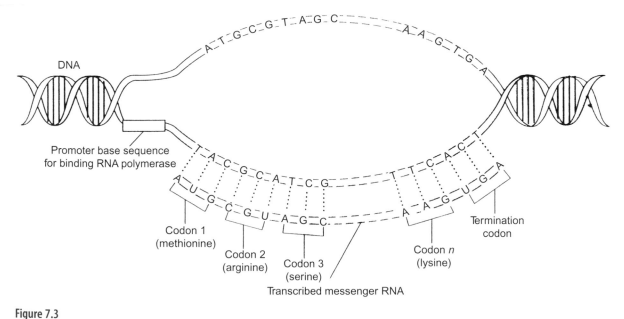

Figure 7.3

In the nucleus of the cell, the genetic information contained in the sequence of base pairs of one strand of the DNA molecule is transcribed into a complementary strand of messenger RNA (mRNA) under the control of mRNA polymerase.

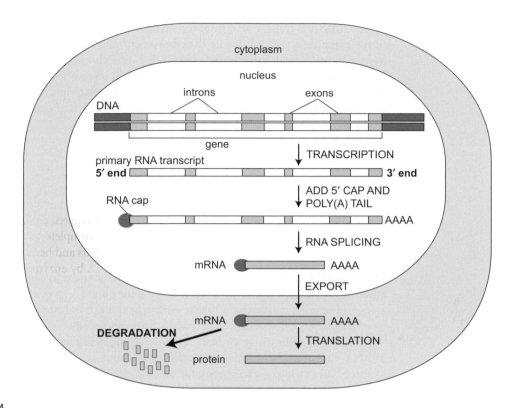

Figure 7.4

Summary of the events leading from gene to protein. Although these events are depicted as occurring one at a time, in a sequence, in reality they often occur simultaneously. For example, the RNA cap is typically added and splicing typically begins before the primary transcript has been completed.

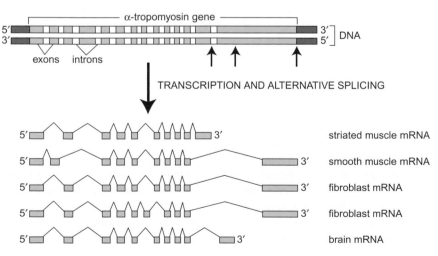

Figure 7.5

Alternative splicing of the α-tropomyosin gene. The protein α-tropomyosin regulates muscle contraction in muscle cells but different forms of the protein are also found in other cell types. Once the primary transcript is made it can be spliced in different ways to produce distinct mRNAs that give rise to different proteins. The arrows just below the gene indicate the positions where cleavage and polyA addition can occur.

The mRNA is translocated from the nucleus of the cell, where it was formed, to the cytoplasm, which is where the ribosomes—the structures on which proteins are synthesized—are located. Although a molecule of mRNA can be quite large, it can pass through the pores in the nuclear membrane. The process of transcription can be regulated by various hormones, nutrients, and metabolites, either directly or via intracellular signalling pathways. Further detail on this can be found in Chapter 8.

Translation

Translation is the process by which information in the sequence of bases on the mRNA molecule is used to determine the sequence of amino acids in the polypeptide chain that is synthesized

The process of translation allows the information contained in the sequence of bases on the mRNA molecule to be used to determine the sequence of amino acids in the polypeptide chain that is synthesized. Each amino acid is denoted by a specific sequence of three base pairs—the genetic code—with each of these sequences being known as a codon. As there are four different nucleotide bases in RNA—adenine (A), guanine (G), cytosine (C), and uracil (U)—combinations of three bases (triplets) can specify for up to $4^3 = 64$ amino acids. In fact there are only 20 different amino acids used in the synthesis of proteins and so the genetic code is said to be 'degenerate', which means that there is more than one codon for each amino acid. For example, the amino acid alanine is coded for by the triplets GCU, GCC, GCA, and GCG; lysine is coded for by the triplets AAA and AAG; and methionine is only coded for by the triplet AUG. Certain codons (UAA, UGA, and UAG) act as a code for chain termination, signalling the end of translation of the mRNA information into a polypeptide chain.

Transfer RNA molecules carry specific amino acids to the ribosomes where protein synthesis takes place

Transfer RNA molecules (tRNA) are found in the cytoplasm and as you can see in Figure 7.6 they have a clover leaf structure. tRNA molecules contain one specific binding site (an anticodon) that recognizes and binds to the codon and another that binds the appropriate amino acid. Ribosomes contain binding sites for two tRNA molecules and a site just below these along which the mRNA strand can progress. The amino acids are thus brought into proximity and form peptide bonds in the appropriate sequence (Figure 7.6). The process is initiated when the first tRNA molecule together with its bound amino acid is positioned on the mRNA: this first amino acid is always methionine, and the rate at which this initiation step occurs is probably crucial in the overall control of the rate of protein synthesis. Elongation of the peptide chain is terminated when a sequence

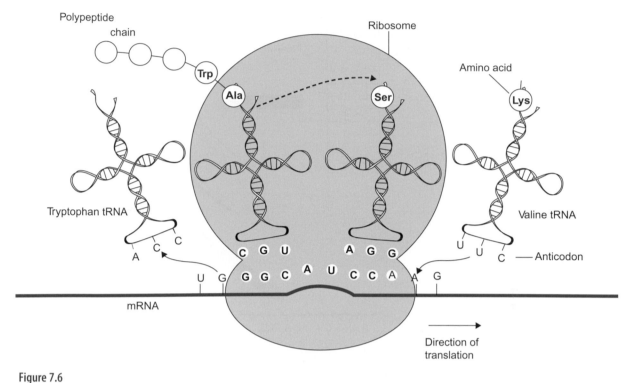

Figure 7.6

On the ribosome, the information carried in the messenger RNA (mRNA) molecules is translated into a specific sequence of amino acids that are positioned by transfer RNA (tRNA) molecules with the appropriate anticodons.

of codons that do not correspond to any of the amino acids is encountered. In this way, the sequence of bases in the mRNA determines the sequence of amino acids in the protein (a simplified overview of the transcription and translation processes is shown in Figure 7.7), which in turn determines how the protein will fold (i.e. its three-dimensional or tertiary structure). The three-dimensional structure of a protein directly determines its function (see Chapter 2). It is this shape and the surface presentation of amino acids that enable highly specific lock-and-key recognition between protein partners and metabolites, hormones, and other signal molecules in the communication circuit, as well as substrate specificity and enzyme activity. There are several means of modifying three-dimensional protein structure following the translation process (this is known as post-translational modification). These processes include phosphorylation, cleavage of peptide sequences, glycosylation (covalent addition of a carbohydrate molecule), or lipidation (covalent addition of a lipid molecule). These may be used to alter specificity or affinity of binding to substrates and enzyme activity.

The length of time that a mRNA molecule persists in the cell affects the amount of protein produced from it, since the same mRNA molecule can be translated many times. Each mRNA is eventually degraded into nucleotides but

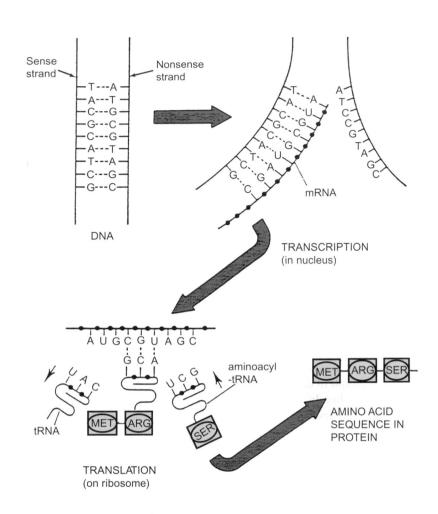

the lifetimes of mRNA molecules differ considerably with some lasting more than 10 h and others lasting less than 30 min. In general, proteins present in high amounts are translated from mRNAs that have long lifetimes, whereas those proteins present at low levels, or whose levels must change rapidly in response to signals, are typically synthesized from short-lived mRNAs.

Gene expression

The partial expression of the genetic information distinguishes one cell type from another

Each cell in the human body contains all of the genetic information necessary to make all of the other cells, but this information remains repressed, and it is this partial expression of the genetic information that distinguishes a muscle cell from a kidney cell. This implies that there is a unique, tissue-specific DNA sequence in or around certain genes expressed in a specific tissue, such as skeletal muscle. A unique DNA sequence that is present only in genes for some of the muscle-specific proteins has been found and appears to be the major component responsible for muscle-specific transcriptional stimulation of several contractile proteins including actin, myosin heavy chain, and troponin. Further research is required to determine whether the coordinated expression of a subset of muscle genes, which respond to exercise training, may occur by means of recognition of a consensus DNA sequence by some kind of 'exercise signal' consisting of a single protein or single oligonucleotide.

Exercise training modifies gene expression in muscle

The coordinated upregulation of sets of muscle proteins during skeletal muscle adaptation to exercise training is well known. For example, heavy resistance training results in a parallel increase of most proteins of skeletal muscle, which is very similar to the parallel upregulation of muscle-specific proteins that occurs as myoblasts develop into myotubes during embryonic development. An attractive hypothesis is that a common set of DNA regulatory sequences would be used by muscle-specific proteins in development and in the hypertrophy that occurs in adult skeletal muscle in response to a weight-training programme.

Another example of differential gene expression during exercise training is the increase in mitochondrial density without a change in muscle size that occurs as a result of endurance training. In this situation it is unlikely that the 'exercise signal' would interact with the same DNA regulatory sequence controlling contractile protein gene transcription. Rather, the endurance exercise factor would interact with a consensus DNA sequence found uniquely in the regulatory region of the mitochondrial genes. The adaptations that occur in muscle with training reflect a change in the expression of the genetic material: endurance training results in an increase in the rate of synthesis of the oxidative enzymes. The control of protein synthesis and the expression of the genetic material can be achieved in a number of different ways.

Transcriptional control alters the concentration of mRNA, and this may be particularly important in the liver where a number of proteins have short half-lives. Control is achieved by regulation of the activity of the mRNA polymerase. Repressor proteins, which are activated or inhibited depending on the availability of specific substrates, allow this control to be exerted. Several hormones exert their effects in this way. The hormone (or a hormone–receptor complex) may bind to a region of DNA on a sensor gene, causing the transcription of an adjacent integrator gene, and resulting in the production of an activator strand of RNA, as illustrated in Figure 7.8. The activator RNA then binds to a receptor gene, which permits the expression of one or more structural genes that are transcribed into mRNA molecules that leave the nucleus and are translated into structural proteins or enzymes on the ribosomes in the cytoplasm.

Figure 7.8

Hormonal regulation of transcription (gene expression) and, hence, synthesis of specific proteins.

Techniques

Polymerase chain reaction (PCR). In molecular biology, the polymerase chain reaction (PCR) is a technique to replicate or amplify a single or few copies of a piece of DNA across several orders of magnitude, generating thousands to millions of copies of a particular DNA sequence. The method relies on thermal cycling, consisting of cycles of repeated heating and cooling of the reaction for DNA melting and enzymatic replication of the DNA. Primers (short DNA fragments) containing sequences complementary to the target region along with a DNA polymerase (after which the method is named) are key components to enable selective and repeated amplification. As PCR progresses, the DNA generated is itself used as a template for replication, setting in motion a chain reaction in which the DNA template is exponentially amplified. PCR can be extensively modified to perform a wide array of genetic manipulations. Almost all PCR applications employ a heat-stable DNA polymerase, such as Taq polymerase, an enzyme originally isolated from the bacterium *Thermus aquaticus*. This DNA polymerase enzymatically assembles a new DNA strand from DNA building blocks, the nucleotides, by using single-stranded DNA as a template and DNA oligonucleotides (also called DNA primers), which are required for initiation of DNA synthesis. The vast majority of PCR methods use thermal cycling: alternately heating and cooling the PCR sample in a defined series of temperature steps. These thermal cycling steps are necessary first to physically separate the two strands in a DNA double helix at a high temperature in a process called DNA melting. At a lower temperature, each strand is then used as the template in DNA synthesis by the DNA polymerase to selectively amplify the target DNA. The selectivity of PCR results from the use of primers that are complementary to the DNA region targeted for amplification under specific thermal cycling conditions. Developed in 1984 by Kary Mullis, PCR is now a common and often indispensable technique used in medical and biological research labs for a variety of applications including the identification of genetic fingerprints (used in forensic sciences and paternity testing); and the detection and diagnosis of hereditary diseases. In recent years PCR has been used in a variety of studies in human exercise. These include functional analysis of genes and changes in gene expression in response to acute bouts of exercise and periods of exercise training. In particular, the development of assays using RNA isolation techniques and the reverse transcription PCR assay have allowed exercise scientists to quantify levels of mRNA in tissues providing new information on gene activation and transcription in response to acute and chronic exercise and the effects of nutritional interventions on these responses. In 1993 Mullis was awarded the Nobel Prize in Chemistry for his work on PCR.

Translational control occurs at the point of assembly of the amino acids under the control of mRNA without any change in the amount of mRNA. There is some uncertainty as to how this control is achieved, but it probably involves the initiation process.

The regulation of protein expression within cells is of major importance in translating the genome into function. Rates of protein synthesis and degradation, localization of proteins within the cell (e.g. myofibrillar and sarcoplasmic and mitochondrial in skeletal muscle fibres) and outside the cell, phosphorylation, glycosylation, lipidation and cleavage, and the binding of allosteric activators or inhibitors determine the activity of different proteins. Generally, the proteins in the cell do not exist in isolation, but form a protein network that consists of protein interactions and pathways that connect in a specific orchestrated manner. Proteins coalesce into such networks, pathways, and circuits in response to specific stimuli. As the latter may constantly change, newly formed protein networks are not constant but break apart such that the population state of protein networks is constantly changing within each living cell. However, it is possible to map the state of key nodes in known protein networks and relate this back to cell function. Biochemists may measure the ratio between activated (e.g. phosphorylated) and the inactivated (e.g. dephosphorylated) form of the key signal proteins and from this can be estimated the status of a signal node at the time at which such measurements are made. For example, after a bout of resistance exercise, skeletal muscles that were involved in high force contractions may have a higher proportion of signal proteins within pathways that stimulate myofibrillar protein synthesis or activate satellite cell development.

Mutations

Inherited defects occur when the mutations occur in the gametes (sperm or ova)

Mutations occur when a gene is missing or defective, which means that the specific protein that it makes will also be missing or defective. This most commonly occurs as a result of a mistake in the process of DNA replication during cell division or during the development of the gametes. Such errors that occur in somatic cells can lead to various forms of cancer, but these mutations are not hereditary. Inherited defects occur when the mutations occur in the sperm, or the ovum that is fertilized. Mutations occur randomly and spontaneously at a rate of about 1 in 25 000 gametes. The rate of mutation can be increased by certain factors called mutagens. These include chemical compounds structurally similar to the nucleotide bases in DNA, other chemicals that interfere with the process of DNA replication, and ionizing radiation.

Not all mutations are necessarily harmful, and this may be one of the advantages of the degenerate nature of the genetic code. For example, a mutation that caused the substitution of guanine (G) for adenine (A) in the third position of the mRNA codon AAA would produce AAG, but this triplet sequence codes for the same amino acid, lysine. Even where this type of mutation (base substitution) results in the insertion of a wrong amino acid into the polypeptide chain, this may not necessarily affect the structure of the protein sufficiently to prevent its normal function, although it would be more likely to do so if the wrong amino acid was located within the active site of an enzyme.

Genetic polymorphism is the basis of variation from individual to individual. DNA polymorphism is defined as a difference from the generally accepted (normal) gene sequence, and occurs in at least 1% of the general population. Single-nucleotide polymorphisms are the commonest form of genetic variability and are the result of a single nucleotide substitution in a DNA sequence. For example the nucleotide sequence ACGAAA could be replaced by ACGTAA. These single nucleotide polymorphisms occur roughly every 1000–2000 nucleotides in the human genome and to date about 22 000 have been sequenced and their relationship to the gene determined.

More drastic mutations can occur when a nucleotide base in the DNA strand is deleted altogether, or an extra base is added. Both these events result in a nonsense code from the point of the mutation onwards in the DNA strand. Similarly, the wrong insertion of a chain termination code (e.g. AAA coding for lysine becoming UAA) results in premature termination of the polypeptide chain. Defective or missing genes can cause hereditary metabolic disorders. One example is McArdle's disease in which the enzyme glycogen phosphorylase is affected. Sufferers of this condition are unable to break down muscle glycogen, which severely restricts their ability to exercise at anything more than a very moderate intensity. The most common inherited disease is cystic fibrosis in which a chloride channel protein is defective. One in 20 of the human population carries this defective gene.

Principles of heredity

Inheritance of traits

A person inherits two sets of genes controlling every physical trait

In each somatic cell of the body (e.g. from heart, liver, kidney, intestine, etc.) there are 23 pairs of chromosomes, so that each cell contains 46 chromosomes, which is called the diploid number. Each somatic cell contains 22 pairs of autosomal chromosomes and one pair of sex chromosomes (XX = female; XY = male). When these cells divide (a process called mitosis) the genetic material is duplicated just before separation such that the genetic information is passed on unchanged. The production of the gametes (sperm in males and ova in females) involves reduction division (meiosis), which results in the gametes containing 23 chromosomes (the haploid number). At fertilization, these combine to produce a zygote with the full complement of 46 chromosomes (see Figure 7.9). The gender of the offspring depends on whether the sperm that fertilizes the ovum contains an X or a Y sex chromosome, as each ovum only contains one X chromosome and 22 autosomal chromosomes.

A person inherits two sets of genes controlling every physical trait (characteristic or phenotype), one from the mother and one from the father (if these

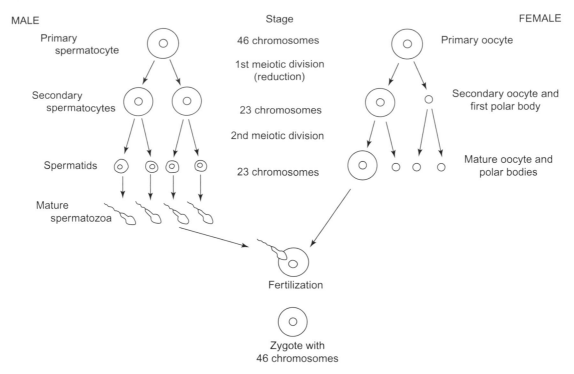

MALE | Stage | FEMALE

Primary spermatocyte — 46 chromosomes — Primary oocyte

1st meiotic division (reduction)

Secondary spermatocytes — 23 chromosomes — Secondary oocyte and first polar body

2nd meiotic division

Spermatids — 23 chromosomes — Mature oocyte and polar bodies

Mature spermatozoa

Fertilization

Zygote with 46 chromosomes

Figure 7.9

The gametes—spermatozoa in males and ova in females—contain 23 chromosomes and are formed by a process that involves reduction division (meiosis) of primary cells, which contain 46 chromosomes. Fertilization of an ovum by a single spermatozoon produces a zygote with 46 chromosomes. This undergoes non-reduction division (mitosis) and differentiation to form the embryo. Segregation and random assortment of chromosomes during meiosis gives $2^{23} = 8\ 388\ 608$ possible combinations of chromosomes.

genes are autosomal—that is, not located on the sex chromosomes). If both genes are identical, the person is said to be homozygous for that trait. A person who is homozygous for normal haemoglobin, for example, has the genotype AA, while a person who is homozygous for abnormal sickle-cell haemoglobin has the genotype SS.

If a person inherits the gene for haemoglobin A from one parent and the gene for haemoglobin S from the other parent, the person is said to be heterozygous for the trait and has the genotype AS (see Figure 7.10a). This person is a carrier of the sickle-cell trait, but does not have sickle-cell disease. Thus, the gene for haemoglobin A is dominant to the gene for haemoglobin S (or the gene for haemoglobin S is recessive to the gene for haemoglobin A). If both parents are carriers with genotype AS, there is a one in four chance that a child from this mating will have the phenotype of sickle-cell disease (genotype SS), as shown in Figure 7.10b. Other examples of autosomal recessive disorders include cystic fibrosis, albinism, and phenylketonuria. These patterns of inheritance were first

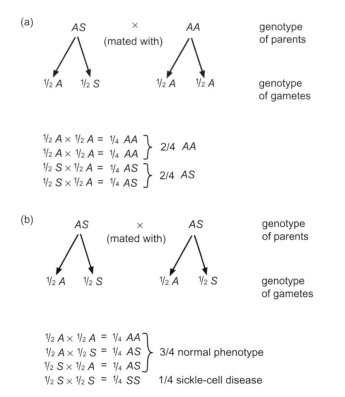

(a)

AS   ×   AA   genotype of parents
(mated with)

½ A   ½ S   ½ A   ½ A   genotype of gametes

½ A × ½ A = ¼ AA
½ A × ½ A = ¼ AA } 2/4 AA
½ S × ½ A = ¼ AS
½ S × ½ A = ¼ AS } 2/4 AS

(b)

AS   ×   AS   genotype of parents
(mated with)

½ A   ½ S   ½ A   ½ S   genotype of gametes

½ A × ½ A = ¼ AA
½ A × ½ S = ¼ AS } 3/4 normal phenotype
½ S × ½ A = ¼ AS
½ S × ½ S = ¼ SS   1/4 sickle-cell disease

Figure 7.10

Examples of autosomal patterns of inheritance. If a person inherits the gene for haemoglobin A from one homozygous parent (genotype AA) and the gene for haemoglobin S from the other parent (heterozygous, genotype AS), the person is said to be heterozygous for the trait and has the genotype AS. This person is a carrier of the sickle-cell trait, but does not have sickle-cell disease because the gene for haemoglobin A is dominant to the gene for haemoglobin S (or the gene for haemoglobin S is recessive to the gene for haemoglobin A). In this mating, half the offspring will have the genotype AA and the other half will be carriers with the genotype AS. If both parents are carriers with genotype AS, there is a one in four chance that a child from this mating will have the phenotype of sickle-cell disease (genotype SS).

described in flowering plants by the monk Gregor Mendel, and hence are sometimes referred to as Mendelian patterns of heredity.

> In sex-linked inheritance, traits are determined by genes located on the X chromosome

In sex-linked inheritance, traits are determined by genes located on the X chromosome and inheritance of these traits follows a different pattern for males from that for females. This is because the male inherits only one X chromosome (and thus only one set of sex-linked traits) from his mother, while the female child inherits an X chromosome from both parents. Examples of sex-linked recessive disorders are haemophilia (due to abnormal blood clotting factor), colour blindness (due to a defective eye cone protein), and Duchenne muscular dystrophy. This last disease, affecting young boys who become severely disabled by the age of 10 years, affects approximately 1 in 3500 live male births. Identification of the abnormal gene on the short arm of the X chromosome in 1979 was followed by the identification of the gene product, the muscle protein dystrophin, in 1987. The gene abnormality is usually a deletion or a point mutation that disrupts the reading frame. There is also a benign form of muscular dystrophy

(Becker's disease) in which the reading frame is maintained and the gene product, although abnormal in some way, remains semi-functional.

The function of dystrophin is still not entirely clear. It is known to be a large cytoskeletal protein located immediately below the plasma membrane of skeletal muscle fibres. Part of the molecule binds to F-actin and other parts are linked to a complex of transmembrane and subsarcolemmal proteins that link, in turn, to the extracellular matrix of the muscle fibre. It has been suggested that dystrophin acts to stabilize the muscle fibre membrane or is involved in the regulation or stabilization of calcium channels in the sarcolemma.

In sex-linked recessive disorders such as Duchenne muscular dystrophy, a normal female may have either the homozygous dominant or the heterozygous (carrier) genotype, while a male must have either the normal or the affected phenotype. For example, if a normal man mates with a woman who is a carrier for Duchenne muscular dystrophy (D is normal, d is abnormal dystrophin), then their respective genotypes are X^DY and X^DX^d. The probability that a child formed from this union will suffer from muscular dystrophy is one-fourth (25%) and the genotype of the affected child can only be X^dY and hence all sufferers of this condition will be male. All female children formed from this union will of course have the normal phenotype, but the probability that a given female child will be a carrier for muscular dystrophy is one-half (50%).

All the above examples are of traits or phenotypes that are determined by a single pair of genes. There are many body characteristics (e.g. height, intelligence) that are determined by multiple genes and cannot be predicted on the simple Mendelian principles described above. There are, however, familial tendencies. For example, if both parents are tall, there is a greater chance of their children becoming tall than if the parents were both short in stature (see Figure 7.11). Parents of successful gymnasts, for example, are generally short in stature, as are the gymnasts themselves.

Figure 7.11

Curve *a* shows the distribution of heights in a normal population of males. Curve *b* shows the distribution of heights of male relatives of men more than 190 cm in height.

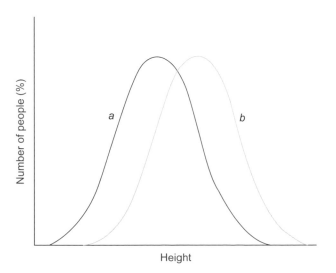

Muscle fibre-type composition: determined by nature or nurture?

In monozygotic twins the muscle fibre-type composition is almost identical

It was stated in Chapter 2 that the fibre-type composition in the major limb muscles is a genetically determined attribute that does not seem to be pliable to a substantial degree by training. In support of this view is the observation that in monozygotic twins the muscle fibre-type compositions have been found to be almost identical. The main limb muscles of non-athletic individuals have, on average, a roughly equal number of Type-I (slow-twitch) and Type-II (fast-twitch) fibres. However, the range of Type-I/Type-II ratios found in members of the general population is fairly large. By contrast, the fibre-type ratios found in elite sprinters are very different from those found in marathon runners and there is limited variation within each group of athletes. For example, the vastus lateralis muscle of successful marathon runners has a high percentage (about 70–90%) of Type-I fibres, while that of elite sprinters contains a higher percentage (about 60–80%) of the Type-II fast-twitch fibres. Among the untrained population, the majority of fast-twitch fibres found in the limb muscles are of the IIX type (glycolytic, fatiguable). In most sports people, the Type-IIa (oxidative-glycolytic, fatigue-resistant) is found in higher proportion than the Type-IIX. This particular difference may be influenced to a significant degree by training. Endurance training increases the oxidative capacity of skeletal muscle without affecting muscle bulk. This increased oxidative capacity results from an increase in the oxidative capacity of all fibre types, not just Type-I. Investigations of human muscle fibre composition among different athletic groups have used the needle biopsy technique to sample muscle. This method usually provides 10–50 mg of muscle tissue containing less than 1000 fibres per sample. A relatively small number of studies of the fibre-type distfibre-type distribution in whole human muscle have been done at autopsy. Regional variations in fibre composition appear to be small, although there is a tendency for the deeper parts of the muscle to contain a higher proportion of Type-I fibres. This suggests that the needle biopsy method is a valid method for estimating the fibre composition of a muscle. Studies of human vastus lateralis (one of the muscles that make up the quadriceps muscle group located at the front of the thigh) indicate that sampling at different sites within the same muscle appears to give no more than about a 6% error (coefficient of variation) in the percentage of the predominant fibre type.

A high degree of heritability has been found for both basic strength and the response to training

Leaders in the field

Bengt Saltin has had an enormous influence on the study of human metabolism and continues to dominate the field almost 50 years after his first publications. He began with some deceptively simple studies on the effects of training and detraining on maximum oxygen uptake in athletes and sedentary individuals that established the heart as the primary limitation to oxygen uptake. He was an early exponent of the muscle biopsy technique and pioneered studies on the metabolic response to high-intensity exercise and to endurance activities. Throughout his career, he has embraced emerging technologies to further our understanding of how the body responds to exercise and other challenges, and he is now working on the interactions between environmental factors and genetic inheritance with a view to understanding how to optimize gene expression. He is also exploring the molecular mechanisms that underpin the associations between inactivity and a range of diseases, including especially diabetes.

So which is more important: nature or nurture? How much of the sporting talent of a great athlete can be attributed to the genes that he or she inherited and how much to the years of practice and physical training? A simple answer is unrealistic. While many of the body characteristics that are often deemed important to athletic capability (e.g. height, body shape, muscle fibre composition, heart size, VO_{2max}) are inborn to a large extent, this always assumes that the genetic potential of the individual is realized through appropriate nutrition and training. Indeed, the responsiveness to training is itself one of the characteristics in which a large influence of the genotype can be demonstrated. In response to an identical training programme, some individuals adapt more—and perform better as a result—than others. For example, the response to strength training varies widely between individuals.

A high degree of heritability has been found for both basic strength and the response to training. A major factor contributing to the increase in strength in response to heavy resistance training is skeletal muscle hypertrophy. A polymorphism of the human angiotensin converting enzyme (ACE) gene has been described in which the deletion (D) rather than insertion (I) allele is associated with higher level of activity of the enzyme in tissues. The D rather than the I variant of the genotype is an important factor in cardiac hypertrophy and has also recently been shown to be involved in skeletal muscle hypertrophy. In response to isometric strength training, gains in strength may be ACE genotype dependent, with some studies showing greater gains in strength shown by subjects with the D allele. Not all studies have shown this, however, even though some have shown an association between the ACE genotype and muscle size. The likely biochemical basis of this is that ACE locally degrades vasodilator kinins and generates a peptide called angiotensin II, which, as well as being a potent vasoconstrictor, is also capable of acting as a tissue-growth modulator. Several studies have identified an association between the presence of a specific allele of the gene encoding the angiotensin-converting enzyme and an individual's endurance or strength capacity, but these association studies do not establish a cause and effect relationship. Many have also used rather small population samples and there is the possibility of a sampling bias in the results of many of these studies. It is unlikely that any single gene will be responsible for a substantial part of individual performance capacity.

Gene doping: a realistic prospect?

Gene therapy, the manipulation of the genetic material to correct disease states, offers the prospect of treating, or even eliminating, some of the major diseases that afflict mankind. Muscle-wasting diseases, including muscular dystrophy, have been the centre of considerable research effort, and the athletic community has not been slow in recognizing the possibilities. Gene doping is the term applied to the use of genetic manipulation techniques to enhance performance in sport.

In the animal world, of course, the genetic component of athletic success is well recognized: this is why millions of dollars may be paid for a young horse that has never raced. Its genetic inheritance conveys a high probability of success, though such success is by no means certain. There have been cases in sport of naturally occurring gene variants in elite athletes that seem to be responsible for outstanding performances. The Finnish cross-country skier, Eero Mäntyranta was found to have an exceptionally high circulating haemoglobin concentration, similar to that achieved with blood doping or with the abuse of Erythropoietin (EPO), the hormone responsible for stimulating red blood cell formation. In his case, this has been attributed to a mutation in the erythropoietin receptor (EPOR) gene, which was identified following a DNA study done on over 200 members of his family. Unfortunately, this finding may not be as simple as it seems, as Mäntyranta failed a drugs test for amphetamines and also later admitted to using hormones that were not specifically prohibited in sport at the time of his best performances. There are other examples of human gene variants, though, including that of a boy born without the gene that codes for myostatin: this protein is produced within skeletal muscle and circulates through the body where it normally acts as a brake on muscle growth, so its absence leads to unusual muscle development. There has also been another recent case of a boy who produces a functional myostatin, but has a defect in his myostatin receptor so that his muscles do not respond to the myostatin signal. A growing awareness of the role of myostatin has led to a search for clinical application of this information and a number of therapeutic options have been explored. Antibodies to myostatin have been developed and shown to have some beneficial effects in mice with muscular dystrophy. There are also some promising early data from studies of compounds that prevent myostatin from interacting with its receptors on the skeletal muscle membrane. None of these avenues has, as yet, led to the development of viable drugs, but these findings have nevertheless provoked great interest from the body-building world. Many products are sold to gullible athletes as 'myostatin-blocking agents' without any evidence of either efficacy or safety. The importance of myostatin has long been known in the farming world, where myostatin-deficient animals have exceptionally large amounts of lean muscle mass. Although beneficial in terms of meat production, however, these animals also have a number of other abnormalities, and this must serve as a warning to any attempts to manipulate muscle growth by altering the myostatin gene.

A number of animal models in which the genetic material has been modified have resulted in animals that have exceptional development of muscle mass or of endurance capacity. Scientists have engineered what they call a 'marathon mouse'. This mouse has been given an enhanced form of the PPAR-Delta gene: this regulates the expression of several other genes and increases the oxidative capacity, and hence fatigue resistance, of the muscles. These mice were able to run about twice as far as normal mice at any given speed even without any endurance training. These animals also had a greatly increased fat oxidation capacity and did not gain weight even when fed a high-fat diet, which has led to suggestions that the same genetic manipulation could be used to bring about weight loss.

Using the animal's own genetic material to produce greater amounts of the proteins that confer these functional abilities means that the proteins are generally not different from the proteins normally occurring in the tissues, making the detection of any manipulation extremely difficult. This is particularly true of the expression of some of the transcription factors that regulate muscle oxidative capacity, contraction speed, and other variables that are important for athletic performance. Because the manipulation occurs within the target tissue, detection would require sampling of these tissues and sampling of muscle would generally be considered too invasive for routine use.

There are some practical issues that have to be overcome before the application of gene therapy becomes a realistic option. As most muscles are involved in activity, making a change in a single muscle would be no use: all muscle involved must be affected, requiring a systemic delivery of the modified genetic material. Modified viruses can be used to carry modified genetic materials into cells: this requires inactivation of some of the virus' own genetic material to prevent replication and hence damage to the host cells, but there is still a possibility of an immune response from the host that has limited most attempts to introduce new genetic materials into cells. Nonetheless, the potential benefits of gene therapy will ensure that these difficulties are overcome.

1. A person's genetic make-up is called his or her genotype. The physical expression of the genotype as particular characteristics or traits (e.g. height, strength, hair colour, etc.) is called the person's phenotype. Success in sport is determined by many factors including motivation, appropriate training, nutrition, and tactics. However, perhaps the most important factor is raw talent in terms of the body's phenotype; in other words the body's physical, physiological, and metabolic characteristics.

2. The genetic material is deoxyribonucleic acid (DNA), which is contained within the nucleus of the cell. The same information is carried by all somatic cells of the body. When cells divide they pass on this information unchanged because the DNA undergoes replication just prior to cell division. Synthesis of proteins is controlled by the genetic information contained in DNA. This determines the sequence of amino acids in a protein chain, as well as initiating and terminating the process. Changes in the body's structure and function are brought about by changing the extent to which the genetic potential is expressed. Although the body's structure is relatively stable, there is a continuous process of protein synthesis and degradation: the half-life of individual proteins varies from less than 1 h to several weeks. This determines the rate of adaptation to environmental stimuli, including exercise training.

3. The parts of DNA that code for specific proteins are called genes. DNA is normally in an uncoiled form and when freed from contact with nuclear protein can be used as a template for RNA and then protein synthesis. Just prior to cell division, DNA condenses into compact forms called chromosomes. There are many genes on one chromosome and in each cell there are 46 chromosomes (23 pairs of chromosomes).

4. DNA consists of two antiparallel chains in a double-helix arrangement. In the uncoiled form, one of these chains can be transcribed to a complementary chain of messenger RNA (mRNA). This step is called transcription and is catalysed by an enzyme RNA polymerase.

5. The entire length of the gene including introns as well as the exons is first transcribed into a long RNA molecule, the primary transcript. After capping and polyadenylation, but before the RNA leaves the nucleus, all of the intron sequences are removed and the exons are joined together. The result is a much shorter RNA molecule containing an uninterrupted coding sequence, This step is called RNA splicing and after it has been completed, the RNA is a functional mRNA molecule that can now leave the nucleus and be translated into protein.

6. The primary transcripts of many genes can be spliced in various ways to produce different mRNAs, depending on the cell type or the stage of development. This process, known as alternative RNA splicing, allows several different proteins to be produced from the same gene.

7. The mRNA passes out of the nucleus through pores in the nuclear membrane and into the cytoplasm where protein synthesis takes place on the ribosomes (which are made of rRNA).

8. The molecule of mRNA attaches to a ribosome. Every three nucleotide bases code for a single amino acid. These base triplets are called codons. Amino acids are carried to the ribosomes by transfer RNA (tRNA) molecules that are specific for a given codon on mRNA and hence specific for one amino acid.

9. Amino acids are joined together on the ribosome in a sequence according to the coding on the mRNA to form a polypeptide chain (protein). This process is called translation. At the end of the sequence, or when a chain terminator codon is reached, the protein molecule leaves the ribosome.

KEY POINTS *(CONTINUED)*

10. Each cell of the body contains 46 chromosomes arranged in 23 pairs; one of each pair is of maternal origin, the other is of paternal origin. Chromosomes are in the nucleus of a cell and are made of DNA. They contain all the genetic information for that individual.

11. Genetic information is contained in genes; these are lengths of DNA along the chromosomes. A gene coding for a particular characteristic is always found in the same position, or locus, on a particular chromosome. Genes code for unique characteristics (e.g. eye colour, height, muscle fibre composition) and shared growth patterns (e.g. *in utero* development, the timing of puberty) and they may also contribute to an individual's basic temperament and intelligence.

12. When a new cell is required for growth, as in foetal development or tissue regeneration after damage, an existing cell divides by mitosis. Just prior to cell division the DNA replicates itself. Each DNA molecule contains one old strand and one new strand. During cell division the duplicate DNA is distributed to two new daughter cells. The result is two identical daughter cells each with 23 pairs of chromosomes (46 in total) carrying a complete set of genes.

13. Female gametes (germ cells) are ova or egg cells; male gametes are sperm. Gametes are formed in the testes or ovaries by a type of cell division that is different from mitosis: it is called meiosis or reduction division. Meiotic division gives rise to germ cells in which the maternal and paternal chromosomes have exchanged some of their genetic material, after which maternal and paternal chromosomes have separated from each other and only one of each pair has entered the new cells. The resulting cells contain only one of each chromosome pair; that is only 23 single chromosomes.

14. When a sperm fuses with an ovum to form the first cell of a new individual, 23 pairs of chromosomes (46 in total) are restored. Each new individual is genetically unique.

15. Many genetically determined features can be modified by the environment in which an individual develops; that is nature (gene expression) may be modified by nurture (environment), e.g. where in the world you were born, how well you have been nourished, or how much training you do.

16. One genetically determined feature that cannot be modified is the gender of an individual. This is dictated by the sex chromosomes X and Y. Sex determination: XX = female; XY = male. The Y chromosome determines maleness. The possible combinations of X and Y from the gametes gives 50% boys and 50% girls.

17. If an error occurs in a gene message it is called a mutation. Mutations are changes in DNA nucleotides. Some changes occur randomly at a low rate or may be induced by certain chemicals (called carcinogens) and radiation.

18. A mutated gene may produce a defective protein and lead to inherited disease. An example of a mutation on a recessive gene is one that can occur on chromosome 7, and if both maternal and paternal chromosomes carry the gene defect the child will suffer from cystic fibrosis.

19. If the mutation is on the X (sex chromosome) it will be carried by a woman (but will be masked by the dominant normal gene) and so it will only appear as a defect in the male, as for example in muscular dystrophy. Colour blindness is another sex-linked condition that almost always occurs in males.

20. If there is a mistake when reduction division takes place, a whole chromosome may become displaced, giving rise to a child with three copies. Sometimes this mistake is lethal and the fertilized egg does not develop, but with chromosome 21 the occurrence of three copies in one individual gives rise to Down's syndrome.

21. In response to an identical training programme, some individuals will adapt more—and perform better as a result—than others. For example, the response to strength training varies widely between individuals. A high degree of heritability has been found for both basic strength and the response to training.

22. Many body characteristics (e.g. height, intelligence) that are determined by multiple genes cannot be predicted on simple Mendelian principles. There are, however, familial tendencies. For example, if both parents are tall, there is a greater chance of their children becoming tall than if both parents are short in stature. Parents of successful gymnasts, for example, are generally short in stature, as are the gymnasts themselves.

23. Some specific genetic mutations confer performance advantages and have led to the prospect of deliberate manipulation to enhance performance. There is no evidence that this is currently practised, but ongoing developments in medical science mean that the opportunities for abuse by athletes will increase in the future.

Selected further reading

Azzazy HME *et al.* (2005). Doping in the recombinant era: Strategies and counterstrategies. *Clinical Biochemistry* 38: 959–965.

Barton ER *et al.* (2002). Muscle-specific expression of insulin-like growth factor 1 counters muscle decline in mdx mice. *Journal of Cell Biology* 157: 137–148.

Brownlee C (2004). Gene doping: Will athletes go for the ultimate high? *Science News* 166: 280.

Charbonneau DE *et al.* (2008). ACE genotype and the muscle hypertrophic and strength responses to strength training. *Medicine and Science in Sports and Exercise* 40: 677–683.

Collins M (ed.) (2009). *Genetics and sports.* Basel: Karger.

de la Chapelle A, Traskelin AL, and Juvonen E (1993). Truncated erythropoietin receptor causes dominantly inherited benign human erythrocytosis. *Proceedings of the National Academy of Sciences USA.* 90:4495–9.

Goldspink G (1999). Molecular mechanisms involved in the determination of muscle fibre mass and phenotype. *Advances in Exercise and Sports Physiology* 5: 27–39.

Komi PV and Karlsson J (1978). Skeletal muscle fibre types, enzyme activities and physical performance in young males and females. *Acta Physiologica Scandinavica* 103: 210–218.

Montgomery HE *et al.* (1999). Angiotensin-converting-enzyme gene insertion/deletion polymorphism and the response to physical training. *Lancet* 353: 541–545.

Murray T (2003). An Olympic tail? *Nature Review Genetics* 4: 494.

Paulaskas A *et al.* (2009). Genetic variability associated with angiotensin converting enzyme (ACE) gene polymorphism in sportsmen pursuing different sports. *Proceedings of the Latvian Academy of Sciences.* Section B, 63: 9–13.

Rivera VM *et al.* (2005). Long-term pharmacologically regulated expression of erythropoietin in patients following AAV-mediated gene transfer. *Blood* 105: 1424–1430.

Schuelke M *et al.* (2004). Myostatin mutation associated with gross muscle hypertrophy in a child. *New England Journal of Medicine* 350: 2682–2688.

Skipper M (2004). Gene doping: A new threat for the Olympics? *Nature Reviews Genetics* 5: 720.

Spurway N and Wackerhage H (ed.) (2006). *Genetics and molecular biology of muscle adaptation.* Edinburgh: Elsevier.

Thomis MA *et al.* (1998). Strength training: importance of genetic factors. *Medicine and Science in Sports and Exercise* 30: 724–731.

Wackerhage H and Woods NM (2002). Exercise-induced signal transduction and gene regulation in skeletal muscle. *Journal of Sports Science and Medicine* 1: 103–114.

Walton JN, Karpati G, and Hilton-Jones J (ed.) (1994). *Disorders of voluntary muscle.* Edinburgh: Churchill-Livingstone.

Wang YX *et al.* (2004). Regulation of muscle fiber type and running endurance, by PPAR-δ. *PLoS Biology* 2: e294.

8 Adaptations to training

Learning objectives

After studying this chapter, you should be able to …

1. appreciate the principles of training to improve performance in sport;
2. describe the principles of training for strength gains;
3. describe the effects of training for speed;
4. describe the effects of training to increase anaerobic capacity;
5. describe the effects of training to increase endurance;

6. explain the nature of the stimuli that induce adaptation to training and the molecular mechanisms that allow adaptation to occur;

7. explain the role of intracellular signalling cascades in bringing about changes in gene expression;

8. appreciate the effects of overreaching and the danger of developing an overtraining syndrome;

9. describe the effects of prolonged exercise and heavy training on immune function;

10. appreciate that diet can influence gene activation in response to exercise and so may modify adaptations to training;

11. understand the sources of free-radical generation during exercise and how training alters the body's antioxidant capacity.

Introduction

All individuals can improve their performance by training

The level of success that an individual can achieve is limited by their genetic potential, and this sets an upper ceiling to performance. To reach that potential, however, requires a sustained period of intensive training, at least for most athletes. There are some sports, usually ball games involving a high degree of skill, where the exceptionally gifted individual can succeed without appearing to pay much attention to training. These are increasingly rare exceptions, and they usually have rather short careers. In the sports that demand strength, speed, or stamina, such individuals are almost unknown. Equally, however, there are those who train to the limits of what they can endure for apparently little return. Whatever the starting point, however, all individuals can improve their performance by training, and the aim of any training programme should be to provide the maximum return in terms of performance gains. Despite the tendency for athletes to attempt to copy the details of the training programmes of their sporting heroes, this is seldom successful. This is in part because of a lack of the physiological, biochemical, and psychological characteristics that are prerequisites for success, but also in part because there is a great variation between individuals in their trainability. This is another characteristic—with a genetic basis—that is often ignored, but it is increasingly being realized that the capacity of an individual to respond to a given training load with favourable adaptations is a vital component of the make-up of the champion.

The type of training, and the frequency, intensity, and duration of training sessions, will vary greatly between sports, and will also be different for different individuals. The balance between training, competition, and rest also varies greatly between sports. Top-level cyclists may race on 100 days per year, and in basketball, baseball, and some other sports players may have several games each week. This offers little opportunity for training between competitions, and a main function of training sessions is to promote recovery. In other sports, such as triathlon or marathon running, there may be no more than a handful of serious competitions each year, with a large part of the year devoted to preparation. This allows the training programme to be planned in advance, with a progression through the competitive season to ensure peak fitness for the major event.

The key elements of a training programme are the intensity, duration, and frequency of the training stimulus that is applied. It is axiomatic that the response of the organism is specific to the stimulus applied. It is possible, for example, to train only one limb using a weight-training programme designed to increase strength and muscle mass. If the training is sufficiently intense and prolonged, there will indeed be an increase in the size of the trained muscles and in the load that can be lifted. The other arm, which has not been trained, will not show any increase in size, but it may show some strength gains, indicating that there may be changes in central nervous system control of the activation of the muscles that responds to training. Generally, however, there is little cross-over of training effects. Figure 8.1 shows the effects of training one leg for strength and the other leg for endurance: the strength of the strength-trained leg increases progressively over the course of the training programme, but there is no gain in strength in the endurance-trained leg. The endurance-trained leg will show an increased endurance capacity, along with the accompanying biochemical adaptations: there will be an increase in mitochondrial size and number, an increase in

Figure 8.1

Changes in muscle strength in response to an 8-week training programme. One leg was trained for strength while the other leg was trained for endurance. Only the strength-trained leg showed a gain in strength.

oxidative enzyme activity, and an increase in capillary density along with other adaptations. These changes reflect changes in the rates of synthesis and breakdown of specific proteins in the trained muscles.

These complexities give an infinite variety of combinations in any training programme, and the coaching of athletes is therefore often described as an art rather than a science. However, there is a strong scientific basis on which a training schedule can be built, and an understanding of this will help to optimize any training programme. There are, however, also serious limitations to the scientific studies of training that have been carried out and the serious athlete might do better to consult an experienced and well-qualified coach than to seek the advice of a scientist whose experience of sport may be derived only from the laboratory.

Training strategies and the associated adaptations

The principles of overload, specificity, and reversibility govern the nature and extent of physiological and metabolic adaptations to training

Both physiological and biochemical adaptations occur as a result of the performance of repeated bouts of exercise over several days, weeks, or months. These adaptations improve performance in specific tasks. The nature and magnitude of the adaptive response is dependent on the intensity and duration of exercise bouts, the mode of training, and the frequency of repetition of the activity, genetic limitations, and the level of prior activity of the individual.

In order to bring about effective adaptation, a specific and repeated exercise overload must be applied. A general principle is that adaptation to training will only occur if the individual exercises at a level above the normal habitual level of activity on a frequent basis. The appropriate overload for any person can be achieved by manipulating combinations of training intensity, duration, frequency, and mode. Another important principle is that physiological and metabolic adaptations to training are generally specific to the nature of the exercise overload. Training for speed and strength induces adaptations that are different from those elicited by endurance training. The major effects of endurance training on skeletal muscle are on its oxidative capacity and its capillary supply. Strength training, however, mainly influences the size (cross-sectional area) of a muscle and thus its force-generating capacity. The specific exercise mode is also important; for example, the development of endurance capacity for running, cycling, swimming, or rowing is most effectively achieved when training involves the specific muscles to be used in the desired activity. This is because regular exercise induces adaptations that are both central (e.g. improvements in cardiac performance) and peripheral (e.g. improvements in local muscular performance).

Techniques

The Enzyme-Linked Immunosorbant Assay (ELISA). The enzyme-linked immunosorbant assay (ELISA) method allows researchers to quantify the concentration of an enormous range of molecules, even those like hormones, cytokines, and growth factors that may be present in very low (nanomolar) concentrations in the tissues and body fluids. The ELISA technique is based on a sandwich principle. First, a 96-well plate is coated with a specific capture antibody against the molecule of interest (all commercially available ELISA plates come with the individual wells pre-coated with the specific capture antibody). Next, samples of interest, standards, and control samples are added to individual wells and the plate is incubated for a fixed period of time during which the molecule of interest is captured by the specific capture antibody. Following the incubation period the plate is washed several times to remove any unbound molecules from the wells of the plate. Next, a secondary antibody that is conjugated to an enzyme (e.g. peroxidase) is added to each well. Importantly, this secondary antibody recognizes a different epitope (an epitope is a site on a large molecule against which antibodies will be produced and to which the antibody will bind) on the molecule of interest to that recognized by the capture antibody. During a second incubation period, the enzyme-conjugated secondary antibody binds to the molecule of interest that has bound to the capture antibody, thus completing the 'sandwich'. Following another series of washes to remove any unbound secondary antibody, a chromogenic substrate of the enzyme is added to each well. The reaction of the enzyme with its substrate converts the substrate to a product that produces a colour change. Thus, the greater the amount of enzyme present (i.e. the greater the amount of molecule of interest bound to the capture antibody), the greater the colour development. To quantify the amount of the molecule of interest present within each well the absorbance (also known as the optical density or extinction) is determined at a specific wavelength on a microplate reader (a fiberoptic multichannel spectrophotometer). Next, a standard curve is generated via a set of standards of known concentration and the concentration of the molecule of interest present in each well is interpolated from the calibration curve. While all ELISAs follow a similar set of procedures and are based on the same principles as described above, variations on this theme are common. The ELISA is an extremely sensitive, specific, and versatile tool, and can be used for the measurement of many biological molecules, e.g. cytokines, hormones, adhesion molecules, soluble receptors, intracellular signalling proteins, and mRNA.

Adaptations to training are essentially transient and reversible: after only a few days of detraining, significant reductions in both metabolic and work capacity are demonstrable and many of the training improvements are lost within several months of stopping regular exercise. Excessive training loads may cause breakdown and loss of performance, a condition referred to as 'overtraining syndrome'. Sufficient time for regenerative recovery is required within a training programme to allow morphological adaptations to occur. Muscle is an extremely plastic tissue, and although genetic factors are the major determinant of the quantity and quality of muscle present in an untrained individual, considerable changes in the functional characteristics, morphology, and metabolic capacity can be induced by training.

Training for strength

Training with weights can increase the force-generating capacity of the individual muscle fibres and cause them to increase in size

Strength training is used by many athletes, not only in sports where peak force generation is important, but also where high power must be generated. The work done is the product of force and the distance through which it acts, and power is the rate at which work is done. Few sports call for the application of force without movement, but there are situations, as for example in a rugby scrum, where very high forces are generated without any external work being done (i.e. the distance moved is zero).

Training with weights can increase the force-generating capacity of the individual fibres that make up a muscle, and can also cause the fibres to increase in size (hypertrophy). By choosing a suitable training programme, these different aspects can be emphasized. Many athletes want both an increased strength or power and an increased muscle mass: this is important in many team games such as football and rugby, where a high body mass is a distinct advantage. In some sports, however, including not only weightlifting but also boxing, wrestling, judo, and other martial arts, competition is by body weight category. These sports are divided into weight categories because of the great advantage conferred by a high body mass, and the recognition that a small individual is unlikely to beat a much larger competitor. Nonetheless, speed, strength, and power are important in these sports, and the training programme must be chosen carefully to maximize gains in these areas without increasing muscle mass so much that the competitor moves up to a higher weight category. For the bodybuilder, the opposite is true: there are no prizes in bodybuilding for being strong, and success depends instead on a large muscle mass, uniform muscular development, and a very low level of subcutaneous fat to allow maximum muscle definition.

The intensity and duration of training sessions will largely determine the training response

To some degree, all of these characteristics—strength, power, and muscle mass—can be developed selectively by a suitable choice of training programme. The specificity is not absolute, and some overlap will occur, but the strength athlete who competes in a specified weight category must be careful not to gain so much extra muscle mass that he moves up a weight class. At the elite level in weightlifting, all competitors have a very low level of body fat—with the exception of those in the super-heavy-weight class, where there is no penalty for a high body fat content—and all have spent many years in rehearsing and refining their technique. Strength, measured as the greatest weight that can be lifted, is closely related to body mass (see Figure 8.2a), but strength per unit body mass decreases as the body mass increases (as illustrated in Figure 8.2b). Actin and myosin are the proteins in muscle that interact to generate force, so strength-training programmes are aimed at increasing the muscle content of these proteins rather that simply increasing total muscle mass.

As with all training programmes, the intensity and duration of training sessions will largely determine the training response. Intensity is critical, as this will dictate the extent to which the different muscle fibres are recruited and therefore subjected to a training stimulus. If the force that a muscle is asked to produce is low, this can be achieved without the need to activate all of the muscle fibres.

Figure 8.2

(a) Winning performance in an Olympic competition (the 1976 Games in Montreal) plotted against body weight classes (body mass on horizontal axis). There is a very strong linear relationship. A similar plot can be obtained for other championship events or by plotting world records for each weight class. (b) Effects of correcting for body mass on the relationship between strength and weight class. The individuals with the lowest mass have the greatest strength per unit mass.

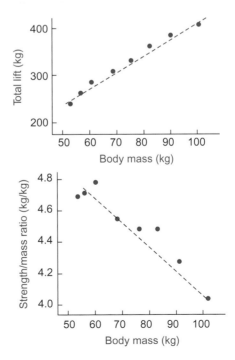

Because of the way the neural control of contraction is organized within the central nervous system, the same fibres are always activated first. These are the slow-twitch Type-I fibres, which have a high oxidative capacity, a low anaerobic capacity, and a high resistance to fatigue. The nerves that control the fast-twitch Type-II fibres are activated only if:

- the load exceeds that which can be lifted by the other (Type-I) fibres; or
- the velocity of movement is very high; or
- other fibres have been fatigued or damaged by prior exercise.

Just as training induces changes only in those muscles that are trained, so any muscle fibre that is not involved in force production will not experience a training stimulus. If the aim is to train the whole muscle, training must involve either high loads that will fully activate all of the muscle fibres or lighter forces that are repeated until fatigue of the slow-twitch fibres forces recruitment of the fast-twitch fibres to occur. If hypertrophy is to occur, the rate of protein synthesis must exceed the rate of degradation. This can be achieved by an increase in the rate of synthesis, a reduction in the rate of degradation, or a combination of both.

Increases in muscle size are largely the result of increases in the cross-sectional area of fibres

The number of muscle fibres is fixed very early in life and even the most intense weight-training programme does not seem to be able to induce growth of new muscle fibres. There has been some speculation as to whether increases in muscle size are the result of increases in the cross-sectional area of fibres alone (the process of muscle hypertrophy), or whether there is the possibility of an increase in the number of fibres (hyperplasia) as a result of the splitting of existing fibres. It has been suggested that once a fibre has increased in size (i.e. diameter) beyond a certain point, the internal stresses that result when that fibre is active will cause it to spit into two smaller fibres. The evidence for this is not strong, however, and results from animal models that involved the separation and counting of all the fibres in muscles from trained and untrained limbs have not provided evidence of an increase in fibre number.

Responses begin within a few training sessions, and there are clearly adaptations occurring within the central nervous system that allow increased force production before there is any significant change in muscle fibre size. However, there are also changes to the contractile proteins within only a few training sessions. Changes in the distribution of the myosin heavy chains—perhaps reflecting interconversion of muscle fibre subtypes—have been seen after only a few training sessions. Measurable increases in muscle size are seen after about 15 training sessions, but this probably reflects the relative insensitivity of the methods used to measure muscle size. Older studies relied on simple anthropometric techniques. Typically, the girth of a limb would be measured, and an estimate would be made of the thickness of the skin and the subcutaneous fat layer at that site using callipers. This then allows calculation of the area occupied by muscle and bone. There are, of course, serious limitations to this method, apart from the limited precision of the measurements. When applied, for example, to the thigh, the calculated area includes not only the knee-extensor muscles at the front of the thigh but also the hamstring muscles at the back of the leg. Bone and some fatty tissue are also included. More recently, developments in medical imaging techniques have allowed cross-sectional images through the body to be obtained by computed tomography (CT, an X-ray technique) and by magnetic resonance imaging (MRI), which relies on the chemical composition of the tissues. Examples of CT

Figure 8.3

Examples of computed tomography (CT) scans taken at mid-thigh level: (a) untrained female; (b) male body builder.

scans taken at mid-thigh level are shown in Figure 8.3. Even with these improved methods, however, the rate of protein synthesis in muscle tissue is too slow for small changes in muscle mass in response to training to be measured.

Training for speed

> The sprinter's training programme is designed to enhance muscle mass, reaction time, and anaerobic capacity

The successful sprinter is characterized by certain well-defined physiological and biochemical characteristics. These include a large muscle mass, with a high proportion of that in the form of fast-twitch (Type-II) muscle fibres, a fast reaction time, and a high capacity for anaerobic energy metabolism. The sprinter's training programme is designed to enhance those characteristics, and involves a large amount of strength training using heavy weights, in addition to specific sprint training.

> Sprint training does not increase the concentration of ATP or PCr in the muscle

The increased muscle mass means a greater store of phosphocreatine (PCr), a greater capacity for anaerobic glycolysis, and a greater buffering capacity. There is no change in the concentration of ATP or PCr in the muscle, but greater power is generated because of the greater muscle mass. In sports where body mass must be moved, this imposes a weight penalty, but the increased muscle mass is clearly a benefit.

Changes in performance capacity would occur simply because of the increase in muscle mass, but there are also specific adaptations taking place. The activity of some of the key glycolytic enzymes (e.g. phosphofructokinase) is increased and there is an increase in buffering capacity, even when these are expressed per unit muscle mass. This may be accounted for in part by a selective hypertrophy of the fast-twitch muscle fibres. Even though the muscle composition remains largely unchanged when expressed as the proportions of the different fibre types present, a selective enlargement of the Type-II fibres means that they account for a greater fraction of the total muscle mass. This means that it may be misleading to describe muscle composition in terms of the proportions of the different fibre types present, as is normally done. Consider the sprinter who begins training with 60% fast-twitch fibres and 40% slow-twitch fibres, and with both fibre types of equal average cross-sectional area (5000 μm^2): this is a lower proportion of fast-twitch fibres than is normally found in elite-level sprinters. From this, we might conclude that this individual will not succeed at the highest level as a sprinter and might be forced to move up to the longer distances in the search for success. If, however, in response to training, the Type-II fibres increase in size by 30%, with no change in the size of the Type-I fibres, the proportion of the total muscle represented by fast-twitch muscle is now closer to 70%. A 50% increase in the average size of the Type-II fibres coupled with a 20% decrease in the size of the Type-I fibres will increase the effective proportion of Type-II fibres to close to 75%. Training alone, therefore, cannot achieve the results that come from favourable genetic endowment, but training can still result in a very substantial remodelling of the muscles.

 Sprint training results in changes in the activity of specific muscle enzymes

There are also changes in enzyme activity and these are highly selective depending on the nature of the stimulus applied. There are problems comparing responses of different individuals, even when the training programme is standardized. Genetic potential, previous training history, motivation, diet, and a whole host of other factors are all different and may affect the outcome. In the laboratory, some of these difficulties may be eliminated by using subjects as their own controls. This is best done by training only one limb or by using a different training programme on each of the limbs. This was done in a study where subjects trained one leg on a cycle ergometer using high-intensity sprints lasting 6 s and trained the other leg with 30-s sprints. Muscle biopsies were taken from each thigh before and after training and showed that the 30-s training resulted in significant increases in some of the enzymes involved in anaerobic metabolism. In particular, the activities of creatine kinase and myokinase both increased by between 10 and 15% after only a few weeks. By contrast, there was no measurable change in the activity of these enzymes in the leg that performed the 6-s sprints.

Other similar studies have involved training one leg for endurance and the other for strength. The specificity of the responses to training is quite clear, even if hardly surprising: endurance training has no effect on muscle strength. Although training studies of this nature are helpful in clarifying both the changes taking place within the tissues and the performance outcomes, they suffer from a shortcoming common to almost all training studies. These studies seldom last for more than a few weeks. It is not surprising that it is difficult to recruit subjects for long-term studies—especially if the end result is one leg that is trained for strength and the other trained for endurance, or if only one leg is trained—and there is a problem in maintaining sufficient motivation to prevent them all from dropping out. This means that most of the information we have from laboratory-based studies of training responses relates only to the first few weeks of a training programme. Athletes recognize that it takes many years of progressive training to succeed at the highest level. Simply comparing elite performers with others from different sports or with a group of sedentary individuals, however, cannot discriminate between training-induced changes and those pre-existing characteristics that perhaps motivated the elite athlete to take part in sport in the first place.

Recognizing the two key issues of high forces being required to increase muscle strength and of training specificity, many sprinters combine these two elements in training. This can be achieved by uphill running, using the athlete's own body mass as resistance. Alternatively, a number of other means of increasing the force required to run can be applied, including towing a heavy object or using a training partner. There is also a need to develop the neural patterns associated with fast running, so there is a fine balance that must be achieved. Some sprinters use downhill running at speeds greater than they can achieve on the flat as a way of attempting to increase running speed, but the effectiveness of this has not been established. Another problem is that downhill running involves a substantial eccentric component and this can result in temporary muscle damage and soreness that may curtail subsequent training sessions.

Training for middle distance: increasing anaerobic capacity

The elite middle-distance runner needs a high capacity for lactate production

It might seem surprising, but one of the important objectives of middle-distance training is to increase the capacity for production of lactic acid, or at least to increase the amount that can be produced before the activity must be stopped. Although there are negative consequences associated with the accumulation of lactic acid in the muscle, the energy that is released by anaerobic glycolysis allows high rates of power to be generated. The first objective is to enhance the rate of glycolysis, which should lead to improvements in performance in situations where the exercise duration is long enough to place significant demands on this energy source but not so long that the limit of the anaerobic capacity is reached. The second objective is to increase the anaerobic capacity, allowing more energy to be produced from this source.

If the rate of glycolysis is to be increased, an increase in the activity of the rate-limiting step is essential. Important enzymes involved in the regulation of the rate of glycolysis are glycogen phosphorylase, phosphofructokinase, and lactate dehydrogenase. The activity of all of these enzymes is reported to increase in response to a training programme consisting of repeated 30-s sprints. There must, however, be some caution in the interpretation of these results. The control of these enzymes is complex, involving the interplay of several different factors. The activity of phosphofructokinase, for example, is inhibited by high levels of ATP and PCr in the cell and is stimulated by high levels of ADP, AMP, and inorganic phosphate. Other activators—which function primarily by releasing the enzyme from the inhibitory effects of ATP—include ammonia, and also the substrate for the reaction, fructose 6-phosphate, and the reaction product, fructose 1,6-diphosphate. Citrate may also play a role: it is certainly an inhibitor of the enzyme in the test tube, but it is less clear that this mechanism is operational in the intact muscle cell. Measuring the peak activity of the enzyme under the artificial conditions of the test tube may give a false impression of its activity under the conditions that prevail in the cell. Because the concentrations of many of the substrates and metabolites in the cell are different at different subcellular locations, it is difficult to reproduce real-life conditions in the test tube.

> Several months of intense anaerobic training can increase the muscle buffer capacity by as much as 40–50%

Notwithstanding these difficulties, however, there is an increase in peak activity of the key enzymes of glycolysis and this is likely to translate to an increased rate of glycolysis. If more lactate is produced, or if it is produced faster, the cell must be able to resist the change in pH that would result if the hydrogen ions were neither removed from the cell nor neutralized within the cell. The buffering capacity of athletes engaged in events where high levels of lactate production are involved is higher than the buffering capacity of sedentary individuals, whereas the muscles of endurance athletes are not different from those of sedentary persons in this regard. An intense anaerobic training programme may increase the muscle buffer capacity by as much as 40–50% within only 8 weeks and this has been found to be associated with an increased capacity to perform high-intensity exercise. Phosphates and histidine-containing peptides (e.g. carnosine) and proteins form the major intracellular buffers in skeletal muscle, while bicarbonate and blood proteins such as albumin (in the plasma) and haemoglobin (in the red blood cells) are important buffers in the extracellular space.

Training for endurance: increasing aerobic capacity

> Cardiovascular and metabolic adaptations to endurance training increase the capacity for fuel oxidation

The primary aim of an endurance-training programme is to enhance the capacity for oxidative metabolism and in particular to increase the oxygen supply to the muscles and to increase the capacity for oxidation of fat. The physiological adaptations that accompany endurance training are well recognized, and the key factors are identified in Table 8.1. An increased cardiac output resulting from an increased stroke volume and an increased perfusion of the muscle resulting from an increase in the density of the capillary network within the trained muscles are perhaps the two most important adaptations.

Because there is a close coupling between the ability of the cardiovascular system to deliver oxygen and the ability of the muscles to use the oxygen that is supplied,

	Rest	**Low intensity exercise***	**Maximal exercise†**
Heart rate	Decrease	Decrease	Slight decrease
Stroke volume	Increase	Increase	Increase
Cardiac output	No change	No change	Increase
Oxygen extraction	No change	No change	Increase
Oxygen uptake	No change	Slight increase‡	Increase
Muscle blood flow	No change	No change	Increase

Table 8.1 Physiological adaptations accompanying a period of endurance training.

* For exercise at the same absolute work rate.
† Exercise at an intensity equivalent to 100% VO_{2max}.
‡ Because of a slightly greater reliance on fat oxidation, which has a ~10% higher oxygen requirement than carbohydrate oxidation.

Figure 8.4

The relationship between the activity of succinate dehydrogenase, one of the enzymes of the TCA cycle, and maximum oxygen uptake in a heterogeneous group of subjects including both sedentary individuals and well-trained athletes.

it is hardly surprising that both cardiovascular function and muscle oxidative capacity are closely linked to performance in endurance events. Figure 8.4 shows the relationship between the activity of succinate dehydrogenase, one of the enzymes of the TCA cycle, and maximum oxygen uptake in a heterogeneous group of subjects including both sedentary individuals and well-trained athletes. Figure 8.5 shows the relationship between cardiac output and oxygen uptake in male and female subjects, again including both trained and untrained subjects. You can see from this figure that there is a good association between oxygen supply (cardiac output) and oxygen uptake. The existence of a correlation, however, does not prove a cause-and-effect relationship.

Muscles adapt to endurance training by increasing the number and size of mitochondria

Figure 8.5

The relationship between cardiac output and oxygen uptake in male and female subjects, again including both trained and untrained subjects. There is a good association between oxygen supply (cardiac output) and oxygen uptake in both submaximal and maximal intensity work.

The biochemical adaptations to a period of endurance training centre around an increase in the capacity for ATP resynthesis by oxidation of fat and carbohydrate. This is achieved by an increase in mitochondrial protein and, more specifically, an increase in the activity of the enzymes and cofactors of oxidative metabolism. There has been much debate as to whether the changes in the central (cardiovascular) or peripheral (biochemical) adaptations are more important and which of these sets the upper limit to the rate of oxidative metabolism. The evidence now indicates quite convincingly that the maximum oxygen uptake is set by the dimensions of the heart, which limits cardiac output. The significance of the increases in oxidative capacity of skeletal muscle in response to endurance training lies in the effect this has on fuel selection.

Key adaptations taking place in the muscles in response to endurance training include:

- an increase in capillary density;
- an increase in myoglobin content (?);
- an increase in activity of oxidative enzymes.

Biopsy samples taken from endurance-trained athletes show that each muscle fibre is surrounded by a greater number of capillaries than are present in the muscles of sedentary individuals. Longitudinal training studies have also shown increases in the number of capillaries after training, even after only a few weeks of training. The major advantage of an increased capillary supply is an increased capacity for the delivery of oxygen to the active muscles. The distance that oxygen has to diffuse from the capillary to reach the deeper regions of the muscle fibres is reduced. There is also an increased capacity for the delivery of fuel—in the form of blood-borne glucose and fatty acids—to the muscle. There is good evidence that the uptake of fatty acids into muscle and their subsequent oxidation is determined primarily by their rate of delivery. Delivery is a function of the plasma concentration, the plasma flow rate, and the surface area available for exchange. Increasing the supply of fatty acids and their contribution to energy supply may be one of the most important adaptations to endurance training.

Myoglobin is an oxygen-binding protein similar in structure to one of the four subunits that make up the structure of haemoglobin. It has a reddish colour and accounts for at least part of the deeper red colour of muscle fibres that have a high oxidative capacity. Myglobin can act as an oxygen store, releasing bound oxygen at times of high demand, but the total amount of oxygen made available in this way is fairly small. A more important function of myoglobin in highly oxidative fibres may be to assist the diffusion of oxygen through the cell. There is evidence of fairly large (approximately twofold) increases in the myoglobin content of animal muscles in response to training, but the evidence that such large increases can occur in human muscle is less convincing.

The effects of training on the oxidative capacity of muscle have been measured in many different ways. In some studies, a piece of muscle is removed and incubated under conditions that stimulate the demand for ATP: the rate at which a range of different substrates can be oxidized is then measured, giving an index of the respiratory capacity of the muscle. These methods are relatively difficult to apply to very small samples of muscle, however, and most studies have looked instead at changes in the activity of one or more of the enzymes involved in oxidative metabolism. Enzymes chosen for this purpose include those involved in the transfer of fatty acids into the mitochondrion (carnitine acetyl transferase), the oxidation of fatty acids (hydroxyacyl-CoA dehydrogenase), the TCA cycle (including succinate dehydrogenase, malate dehydrogenase and citrate synthase), and the respiratory chain (cytochrome c and cytochrome oxidase).

Cross-sectional studies—those where comparisons are made between separate groups of trained and untrained individuals—show that both the maximum respiratory capacity and the activity of most of these enzymes is higher in endurance-trained athletes by a factor of two- to threefold. This type of study also shows a good relationship between performance—measured as the best time for a given race distance—and the activity of these enzymes in the leg muscles. Longitudinal studies—where the same subjects are followed over a period of time during which they follow a prescribed training schedule—show increases in enzyme activity with even small amounts of training, and they also show that the response is generally proportional to the amount of training carried out.

> Is muscle glycogen content affected by training?

Most of the carbohydrate entering the glycolytic pathway during intense exercise is derived from the intracellular glycogen store rather than from blood glucose. There is a finite amount of glycogen in the muscle, and there is good evidence that the availability of glycogen can, in some situations, limit performance in endurance exercise. There have been suggestions that endurance training may increase the amount of glycogen stored in the muscle and that this, coupled with a decreased rate of glycogen depletion, can account for the increased endurance capacity of the trained athlete. It is difficult, however, to be certain that the muscle glycogen content does increase with training. The athlete who trains daily, or even twice a day, undergoes a regular cycle of glycogen depletion and replenishment. If one or two days of rest are taken, together with a high carbohydrate diet, then very high levels of muscle glycogen are achieved, but this is perhaps not, strictly speaking, an adaptation to training.

> Endurance training increases the contribution of fat oxidation to energy supply during exercise and so spares the use of muscle glycogen and blood glucose and decreases the rate of accumulation of lactate during submaximal exercise

There is no doubt that training adaptations in muscle affect substrate utilization. Endurance training increases intramuscular content of triacylglycerol, and increases the capacity to use fat as an energy source during submaximal exercise. Trained subjects also appear to demonstrate an increased reliance on intramuscular triacylglycerol as an energy source during exercise. These effects, and the aforementioned physiological effects of training, including increased maximum cardiac output and VO_{2max}, improved oxygen delivery to working muscle and attenuated hormonal responses to exercise, decrease the rate of utilization of muscle glycogen and blood glucose and decrease the rate of accumulation of lactate during submaximal exercise. These adaptations contribute to the marked improvement in endurance capacity following training.

Alterations in substrate use with endurance training could be due, at least in part, to a lesser degree of disturbance to ATP homeostasis during exercise. With an increased mitochondrial oxidative capacity after training, smaller decreases in ATP and PCr and smaller increases in ADP and P_i are needed during exercise to balance the rate of ATP synthesis with the rate of ATP utilization. In other words, with more mitochondria, the amount of oxygen as well as the ADP and P_i required per mitochondrion will be less after training compared with before training. The smaller increase in ADP concentration would result in less formation of AMP by the myokinase reaction, and also less IMP and ammonia would be formed as a result of AMP deamination. Smaller increases in the concentrations of ADP, AMP, P_i, and ammonia could account for the slower rate of glycolysis and glycogenolysis in trained compared with untrained muscle.

Mechanisms and limitations to adaptation

Two fundamental questions arise from the preceding description of how muscles adapt to training. One is, what is the precise nature of the stimulus (or stimuli) that induce(s) adaptation and the other is what are the molecular mechanisms that allow such adaptation to occur? Answers to these questions are beginning to come from recent advances in molecular biology that are now being applied to skeletal muscle, investigating how this highly adaptable tissue responds to increased use and disuse.

Muscle genes are regulated largely by mechanical and/or metabolic stimuli

It seems certain that the myosin gene family holds the key to muscle adaptability. There are at least seven different versions available, so the muscle fibres are inherently flexible. In theory they can alter their contractile properties by rebuilding their myofibrils using a different type of myosin heavy chain. A fast-twitch muscle fibre could become a slow fibre simply by switching off the gene for the fast myosin heavy chain and switching on the gene for the slow isoform of the protein. Most genes in the cells of the body are switched on and off by the indirect actions of signalling molecules such as hormones or growth factors. Adaptations in muscle in response to training are specific to the muscles used in the activity; unused muscles do not adapt. Thus, it seems that muscle genes are regulated largely by mechanical and/or metabolic stimuli. It is the stretching or contracting of a muscle fibre that turns specific genes on or off. The activity of different myosin genes can be monitored using DNA probes. Animal experiments have elegantly demonstrated that stretch alone and contraction alone (induced by electrical stimulation) in a fast muscle (e.g. tibialis anterior) only slightly affected the activities of myosin genes. However, together these stimuli caused a dramatic shutdown in the synthesis of the fast myosin heavy chain, switching almost exclusively to the slow version of myosin. Immobilization of a slow muscle (soleus) in the rabbit caused the muscle to revert to expressing the fast myosin gene. Thus, it appears that the fast myosin is the 'default option'. A slow muscle such as the soleus needs to be repeatedly stretched to sustain its synthesis of the slow myosin heavy chain. Both these observations are consistent with what we know about the effects of different types of training regimens on human muscles.

Stretching of muscle is a potent stimulus to adaptation

Stretching of the muscle fibres during exercise is one potent stimulus to adaptation. Passive stretch is known to induce muscle enlargement even in the absence of innervation, insulin, growth hormone, or adequate nutrition. The transduction of mechanical forces through the cytoskeleton to the nuclei and polyribosomes of the muscle fibre may occur either directly or via membrane-bound stretch-activated ion channels or stretch-induced alterations in plasma membrane-associated molecules (e.g. mechanoresponsive isoforms of adenylate cyclase). These stretch-related signals could then induce altered gene expression (e.g. of muscle growth factors) and altered rates of protein synthesis and degradation. However, in order for stretch or other stimuli to induce changes in gene expression and rates of protein synthesis there has to be a mechanism linking the signal to its cellular effects. Recent research suggests that mechanical load applied on skeletal muscle activates various signalling pathways leading to enhanced activation of the ribosomal machinery and the selective transcription of genes associated with enzymes or structural proteins. In the past few years we have learnt a lot more about these intracellular signalling cascades.

Activation of an intracellular signalling cascade begins the process of adaptation

The complex process of exercise-induced adaptation in skeletal muscle begins with certain molecular events that trigger an increase in protein synthesis. More specifically, signalling mechanisms triggered by the exercise stress initiate replication of genes (specific sequences of DNA) that enable subsequent transcription into specific mRNA molecules followed by translation of the genetic code into a series of amino acids to synthesize new proteins (see Chapter 2 for a description of the basic processes involved). It is the synthesis of these specific proteins that ultimately results in the metabolic and physiological adaptations. Thus, a small degree of adaptation occurs in response to a single bout of exercise. When the exercise is repeated several times per week, these small adaptations accumulate to produce measureable changes in the metabolic characteristics and/or size of the muscle. The signalling events, and the resulting increases in mRNA, as well as the synthesis of proteins, are dependent on the intensity and duration of exercise, on the type of exercise performed and also, at least to some degree, on the timing and composition of nutrient intake. Clearly, the adaptations to endurance exercise and heavy resistance exercise are quite different. Endurance training results in increased oxidative capacity of skeletal muscle with little or no change in muscle size and strength. The increased oxidative capacity arises largely from an increase in the number and size of mitochondria in the trained muscles. Other metabolic adaptations in muscle in response to endurance training include an increased content of GLUT4 that enables more rapid glucose entry into muscle from blood and greater glycogen storage following prolonged exercise that has depleted the limited stores. In contrast, heavy resistance exercise training results in increased muscle strength, hypertrophy, and increased amounts of myofibrillar protein content but with relatively little change in the activities of the enzymes of energy metabolism. It seems intuitively obvious that these different adaptations to endurance and resistance exercise are likely to be brought about by different signalling mechanisms and pathways.

Most of the key intracellular signalling proteins act as molecular switches

Protein synthesis is regulated by signal-transduction pathways that sense and integrate local and systemic signals and regulate various cellular functions. The main signalling mechanisms are the phosphorylation of serine, threonine, and tyrosine residues by kinases and their dephosphorylation by phosphatases. The growth, metabolism and proliferation of muscle fibres and most other functions are dependent on signals like nutrient availability, pH, partial pressure of oxygen, reactive oxygen species (ROS), and mechanical stimuli. The nervous and endocrine systems provide further inputs. Cells detect this ever-changing mix of signals through specific sensor proteins. Examples of sensor proteins are cell membrane receptors and amino acid, calcium, or AMP sensing proteins. Activation of a sensor protein triggers a cascade of reactions involving the activation of several protein kinases (enzymes that phosphorylate other proteins and by so doing activate or inhibit the activity of these proteins) that provide the link between the signals and the change in cellular function. Signal-transductions pathways control the following processes:

1. transcription of genes (specific sequences of DNA) into mRNA;
2. translation of mRNA into protein;
3. protein modification altering catalytic activity;
4. regulation of protein degradation;
5. regulation of cell division, proliferation, and fusion.

These signal-transduction pathways can then influence a number of cellular events and will ultimately result in altered function. Signalling cascades usually amplify the signal received, making it stronger, so that a few signalling events are enough to evoke a large response. The signalling cascade can also distribute the signal so as to influence several processes in parallel. At any step in the pathway, the signal can diverge and be relayed to a number of different intracellular targets, creating branches in the signal flow diagram and evoking a complex response (Figure 8.6). Furthermore, many steps in the signalling cascade are open to interference (inhibition or amplification) by other factors so that the transmission of the signal can be modulated according to conditions prevailing inside or outside of the cell.

Most of the key intracellular signalling proteins act as molecular switches. The receipt of a signal switches them from an inactive to an active state and they persist in that active state until some other process switches them off. The latter is equally important because if the signalling pathway is to recover after transmitting a signal and make itself ready to transmit another, every molecular switch must be reset to its original unstimulated state. Proteins that can act as molecular switches mostly consist of proteins that can be turned on or off by phosphorylation: the switch is thrown in one direction by a protein kinase (which adds a phosphate to the switch protein) and in the other direction by a protein phosphatase (which removes the phosphate from the switch protein). Many of the switch proteins controlled by phosphorylation are themselves protein kinases and these are often organized into phosphorylation signalling sequences or cascades: one protein kinase, activated by phosphorylation, phosphorylates the next protein kinase in the sequence, and so on, transmitting the signal onwards and, in the process, amplifying it, distributing it, and modulating it.

Figure 8.6

A simplified illustration of intracellular signal-transduction pathways in relation to exercise. Muscle contraction may result in a disturbance of homeostasis. Mechanical stress (stretch, tension), Ca^{2+} and the accumulation of AMP or metabolites are primary signals that affect the phosphorylation state of various kinases. These kinases can influence the transcription process, the transport of mRNA out of the cell, and translation. Hormones and other ligands can influence these signals. The changes in gene expression can result in different adaptations.

The above general description of signalling pathways is very general and probably applies to most cells in most situations. The details of these signalling cascades, the triggers, the proteins, and the kinases involved are different in different tissues and different situations. To understand adaptations in skeletal muscle cells in response to exercise training, we need to know more details about the signalling pathways in muscle.

Disturbances in cellular homeostasis initiate signalling cascades that lead to adaptation in skeletal muscle

 Leaders in the field

Dave Costill is one of those scientists whose career has been driven by a fundamental curiosity about how things work. From the 1960s through to the present day, this has led him to study many different aspects of exercise physiology, but his research has centred on the factors that determine exercise performance and on the adaptations that take place when an individual undertakes a training programme. Unlike many exercise scientists, Dave welcomed elite athletes into his laboratory and his published papers include data collected from some of the world's greatest runners. As an elite masters swimmer himself, he never forgot the practical implications of his research.

During exercise cellular homeostasis is disturbed and will result in the beginning of a cascade of events that ultimately result in an adaptation that will result in less disturbance of homeostasis the next time this same exercise is performed. During exercise these signals include changes in muscle stretch or tension, changes in intracellular calcium ion (Ca^{2+}) concentrations, changes in the energy charge of the cell (i.e. the $[ATP]/[ADP+AMP+P_i]$ ratio), and changes of the redox potential. These signals can be altered by certain hormones and other ligands that can bind to receptors on the cell surface. These primary messengers may then trigger a series of secondary molecular events that increase or decrease transcription or translation as shown in Figure 8.6. The various metabolic or mechanical changes in the muscle that may initiate a signalling cascade are described below.

Mechanical perturbations of skeletal muscle cells (stretch or altered tension) cause activation of a number of signalling pathways including activation of calcineurin, mitogen-activated protein kinase (MAPK), and insulin-like growth factor (IGF) signalling cascades.

Neural activation of skeletal muscle generates an action potential that results in Ca^{2+} release from the T-tubules of the sarcoplasmic reticulum. When the exercise is stopped the Ca^{2+} will return back from the cytoplasm to the sarcoplasmic reticulum. The fluctuations in Ca^{2+} concentration or the release and reuptake of Ca^{2+} in the sarcoplasmic reticulum are different for different types of activities and therefore this could at least partly explain the differences in the adaptive response to exercise. Endurance exercise, for example, would result in more prolonged moderately elevated concentrations, whereas high-intensity exercise would cause shorter periods of very high Ca^{2+} concentrations. Elevations of Ca^{2+} concentration in the muscle cytoplasm will activate calmodulin kinase (CaMK) and calcineurin (CaN).

During muscle contraction ATP is broken down to provide energy. In this process ADP and P_i are formed. ADP is then resynthesized to ATP via glycolysis or oxidative phosphorylation. Some ADP is further broken down to AMP. The ratio of metabolites ADP, AMP, and P_i relative to that of ATP is often referred to as the energy charge. If there is a lot of ATP and little of the metabolites, the energy charge is high. If the concentration of metabolites rises, the energy charge is low. These metabolites are important regulators of metabolism but they also serve as signalling molecules. AMP in particular can activate 5′ adenosine-activated protein kinase (AMPK), a potent secondary messenger. AMPK seems to play a role in the regulation of a variety of processes including glucose uptake, fatty acid oxidation, hypertrophy, and gene expression. The redox potential is another indicator of the energy status of the cell. If the ratio of the oxidized form of the coenzyme nicotinamide adenine dinucleotide (NAD^+) to its reduced form (NADH) is high, this is an indication of poor energy status. The maintenance of redox potential produces highly reactive free-radical oxygen molecules (reactive oxygen species, ROS). These ROS are also thought to play a role in the exercise-induced signalling that will ultimately be responsible for training adaptations. This may work by acting on transcription factors like nuclear factor kappa B (NFκB) and activator protein 1 (AP1).

Hormones can affect the activation of kinases as well. The hormone or other ligand binds to a receptor that then changes the phosphorylation state of a kinase. For example, when thyroid hormone (tri-iodothyronine, T3) binds to its receptor this induces phosphorylation of AMPK in skeletal muscle.

The primary messengers activate a series of secondary messengers. As mentioned earlier, these secondary messengers are often kinases and phosphatases that are activated to pass on the exercise-induced signal. The secondary messengers often involve a complex series of reactions (cascades) that are highly regulated.

> AMPK is responsible for inhibiting anabolic processes that require energy and for stimulating catabolic processes that generate energy

AMPK plays a crucial role in energy metabolism and acts as a metabolic master switch regulating several intracellular systems including the cellular uptake of glucose via upregulation of GLUT4, a sparing of carbohydrate use via an increased expression of pyruvate dehydrogenase kinase (which inactivates pyruvate dehydrogenase, thus limiting pyruvate entry into the mitochondria and reducing glucose oxidation), the β-oxidation of fatty acids, and the biogenesis of mitochondria (Figure 8.7). Overall AMPK is responsible for inhibiting anabolic processes that require energy and for stimulating catabolic processes that generate energy. AMPK possesses an energy-sensing capability that can be attributed to its ability to detect and react to fluctuations in the AMP:ATP ratio that take place during rest and exercise. Muscle contraction is associated with

Figure 8.7

A change in the AMP:ATP ratio will activate AMPK by activating AMPK kinases (LKB1, CaMK kinases, and the TGF-β activating kinase). AMPK then acts as a metabolic master switch, turning on processes that generate ATP and turning off processes that use ATP. It also causes an adaptation so that the next time the same stress is applied the cell will be better adapted.

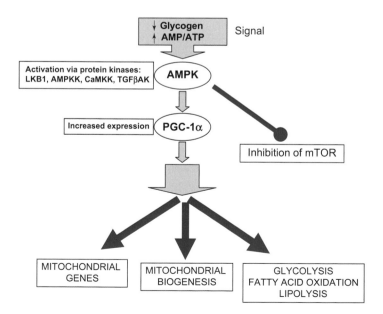

an increase in the demand for cellular ATP, which subsequently increases the AMP:ATP ratio. AMPK is activated by an increase in the AMP:ATP ratio. Acute activation of AMPK will result in a response that aims to conserve energy (limit ATP use) and generate more ATP. For example, AMPK activation will result in an increase in glucose uptake and an increase in fat oxidation so that ATP can be regenerated. When AMP is high and ATP is low, AMP activates AMPK by dislodging ATP from the α-subunit of AMPK and making it more sensitive to phosphorylation by AMPK kinases. In skeletal muscle, the major AMPK kinase appears to be LKB1, although the CaMK kinases and the transforming growth factor (TGF)-β activating kinase may also activate AMPK. Another effect of AMPK activation is the inhibition of protein synthesis via an inhibition of mTOR activity (see below for an explanation of the mTOR signalling pathway).

The peroxisome proliferator-activated receptor γ coactivator family are important in driving mitochondrial biogenesis

One family of transcriptional regulators in particular, the peroxisome proliferator-activated receptor γ coactivator family (PGCs), PGC-1α, PGC-1β, and the PGC-1 related coactivator (PRC), are important in driving mitochondrial biogenesis. The PGCs are coactivators of the peroxisome proliferator-activated receptor γ (PPAR-γ). A transcription coactivator is defined as a protein or protein complex that increases the probability of a gene being transcribed by interacting with transcription factors, but does not itself bind to DNA in a sequence-specific manner. PGC-1α mediates its effects on mitochondrial biogenesis by activating the transcription factors that regulate nuclear genes encoding mitochondrial proteins and that induce expression of mitochondrial transcription factor A, which regulates mitochondrial DNA transcription, thus activating the coordinated expression of mitochondrial proteins. Overexpression of PGC-1α in skeletal muscle and myotubes results in a large increase in functional mitochondria and in increased GLUT4 expression. The activity as well as the expression of PGCs are rapidly increased following a single bout of endurance exercise. PGC mRNA increases up to 10-fold following a single bout of exercise. PGC activation ultimately results in increased oxidative capacity and increased endurance performance and therefore the goal of endurance training is effectively to maximize the activation of PGC signalling pathways in skeletal muscle. AMPK may operate directly upstream of PGCs, regulating the level of PGCs and therefore the metabolic state of the muscle. As a result, training that increases AMPK activity should be beneficial for endurance performance. It has also been suggested that central nervous system activity through β-adrenergic receptors may play a significant role in the activation of PGCs and the subsequent increase in mitochondria. Large rises in FFA have also been suggested as triggers for PPAR-γ activation that may be one mechanism whereby long-term high-fat diets cause an increase in mitochondrial biogenesis.

Activation of the mTOR signalling pathway appears to be crucial in bringing about the increase in muscle protein synthesis following resistance exercise and nutrition

Calmodulin-dependent kinases (CaMK) is a name for a family of different kinases that can detect and respond to changes in calcium concentrations. It has been shown that certain isoforms of CaMK (CaMKII) can also respond to muscle stretch. Once CaMK is activated this will result in the activation of a number of other signalling molecules including nuclear factor of activated T-cells (NFAT) and histone deacetylase (HDAC). Calcineurin is another molecule that is activated by Ca^{2+}. Calcineurin is known to act as a co-regulator of muscle hypertrophy in combination with insulin-like growth factor (IGF) but it also has a role in fibre-type transformation (fast twitch to slow twitch) and the expression of the genes of oxidative enzymes.

The insulin and insulin-like growth-factor pathways play an important role in muscle hypertrophy. Contractile activity stimulates the release of IGF-1 that binds to its receptor that initiates a cascade of molecular events (Figure 8.8). After IGF binds to its receptor insulin receptor substrate 1 (IRS1) is activated and this in turn activates PI3K. The latter activates PDK1 that in turn phosphorylates Akt. Akt has numerous targets including those involved in protein synthesis (mTOR), tuberous sclerosis complex 2 (TSC2) and glycogen synthase kinase 3 (GSK3), protein degradation (Forkhead Box O1 (FoxO1)), glucose

Figure 8.8

The insulin and insulin-like growth-factor pathways play an important role in muscle hypertrophy through the Akt–mTOR signalling cascade.

transport (Akt substrate of 160 kDa; AS160). There is strong evidence that the Akt–mTOR pathway is involved in hypertrophy via activation of translation initiation as well as increasing the ribosomal protein content. The response of Akt to exercise is unclear at the moment and studies have found either increases or no change in response to exercise. This is likely because of the central role of Akt in regulating muscle hypertrophy as well as glucose transport. The Akt response may therefore be highly specific to the type of exercise and is probably influenced by many other factors. mTOR responds to a number of different stimuli (including hormonal, nutrient, and contractile stimuli) and can have effects on mRNA translation, the synthesis of ribosomes, as well as metabolism. Recent studies in humans indicate that skeletal muscle protein synthesis is increased following the ingestion of essential amino acids plus carbohydrates and following a bout of resistance exercise. However, muscle protein synthesis is increased to a greater extent when the nutrients are ingested soon after resistance exercise. Activation of the mTOR signalling pathway appears to be crucial in bringing about the increase in muscle protein synthesis following resistance exercise and nutrition. It is also thought that the direct stimulatory effect of the branched-chain amino acid leucine on muscle protein synthesis is mediated via the mTOR pathway.

Two mTOR protein complexes exist where mTOR binds with a G-β-L protein and either a rapamycin-sensitive raptor or a rictor protein. mTOR-raptor complex is a positive regulator of cell growth, while mTOR-rictor appears to have a key role in Akt activation and actin cytoskeleton regulation. Primary downstream targets of mTOR-raptor include p70 S6 kinase (p70 S6K), eIF4E-binding protein (4E-BP1), and eIF4B. These downstream targets increase mRNA translation and increase protein synthesis. It is generally thought that mTOR plays a central role in the adaptation to resistance exercise but it has also been shown that mTOR promotes mitochondrial gene expression. This seems something of a paradox given that AMPK is important in promoting mitochondria biogenesis but also inhibits the mTOR signalling pathway and thus suppresses protein synthesis. However, it should be borne in mind that the inhibitory effect of AMPK on the mTOR signalling pathway is a rather short-term effect driven by low cell energy status (i.e. in response to fasting or prolonged endurance exercise) and may well rapidly disappear during the recovery phase and with restoration of nutrient availability.

Exercise-induced damage to muscle fibre ultrastructure may be obligatory for satellite cell proliferation and fusion, resulting in hypertrophy

Damage to muscle fibres during exercise may also provide a stimulus for adaptation. High concentrations of a muscle-specific growth factor released from damaged, degenerating (necrotic) muscle fibres, together with the loss of contact inhibition between satellite cells and underlying live muscle fibres, commit satellite cells to proliferate during the first day after muscle injury. Exercise-induced alteration of muscle Z-lines, cytoskeleton, or the extracellular matrix may therefore be obligatory for satellite-cell proliferation and fusion, resulting in hypertrophy. The newly recruited satellite cells must provide the additional nuclei required to maintain nuclear density in hypertrophied muscle. A number of DNA hybridization studies have demonstrated a derepression of a muscle regulatory factor in muscle fibre nuclei during muscle regeneration and a localization of myosin heavy-chain mRNA in focal areas of damage produced in overstretched muscle.

Muscle-specific protein adaptations show a sequential appearance of alterations in groups of proteins

Typically the initial response to exercise is apparent within second or minutes. For example, changes in intracellular calcium are instant. Then, a cascade of reaction follows and the activation of various kinases can take a little longer and some kinases may not reach their maximal activity until several hours after exercise. Gene expression seems to peak between 4 and 12 h after exercise and this may depend on the gene and also the type of exercise performed. Myogenic and metabolic genes appear to peak at 4–8 h after resistance exercise. After endurance exercise, myogenic genes peak after 8–12 h. Thus, the changes in protein synthesis can be observed within hours after exercise but may peak many hours later. In fact, studies have demonstrated increased protein synthesis for up to 48 h after exercise.

Several studies have investigated the signalling pathways that are activated by exercise but to date few studies have linked these changes with actual changes in protein synthesis. It is important to note that although these signals can increase the synthesis of mRNA this does not necessarily mean that protein synthesis is also increased. The rate of protein synthesis is dependent on mRNA degradation, additional translational control mechanisms, transport of mRNA out of the nucleus, and the translation process.

Changes in the expression of the different myosin isoforms (and isoforms of the other contractile proteins) occur during apparent transformations of fibre types induced by cross-innervation of muscles (surgical intervention), chronic electrical stimulation, and disuse atrophy. Levels of thyroid hormones and growth hormone have been shown to influence such changes in myosin heavy-chain isoforms in skeletal muscle experiencing altered contractile activity. Muscle-specific protein adaptations show a sequential appearance of alterations in groups of proteins. For example, specific gene groups appear to respond in the following order: Ca^{2+}-handling proteins change first, followed by mitochondrial proteins, and contractile proteins are altered last. The same order of change in gene sets seems to occur in both the slow-to-fast conversion of postural muscles induced by relief from weight-bearing activity and the conversion of fast-to-slow muscle that can be induced by prolonged electrical stimulation.

Overreaching and overtraining

The overtraining syndrome is a condition in which underperformance is experienced despite continued or even increased training

If the training load is too high with insufficient time for recovery between sessions, the beneficial adaptations that are sought will not occur, and instead there will be a loss of functional capacity. This is often referred to as overtraining, overtraining syndrome, or unexplained underperformance syndrome, a condition in which underperformance is experienced despite continued or even increased training. Although improvements in athletic performance hinge on increasing the training load or overloading, overtraining—a vicious circle of more training producing lower performance and chronic fatigue—seems to be a stress response to training too hard too often, with insufficient recovery time between exercise bouts.

Many different terms have been used to describe the phenomenon of overtraining. These include: overwork, overstraining, overstress, staleness, burnout, chronic fatigue, and overreaching. It is important, however, to distinguish between overtraining, in which there is a chronic decrement in performance that can take weeks or months to recover from, and overreaching, in which a relatively short-term decrement in performance may be experienced, but which is followed within a few days by a full recovery or improvement in performance (supercompensation).

One of the problems is that it is often difficult to distinguish between overreaching and the early stages of overtraining. The difference between overreaching and overtraining is that the athlete recovers within days from the former, whereas overtraining results in sustained reductions in performance and is often (though not always) accompanied by other biochemical, physiological, and psychological changes.

The reasons why some athletes become overtrained while others do not are unclear, although it appears that highly motivated athletes who attempt to sustain very heavy training loads are the most susceptible. An insufficient energy intake during periods of heavy training may be another contributing factor. It is important to recognize, however, that some periods of very intensive training are necessary if peak performance is to be achieved, so simply reducing the training load is not the solution. Other sources of stress, including mental anxiety, may also contribute to the development of overtraining syndrome.

The most common symptoms of overtraining are fatigue, underperformance, difficulty in maintaining the usual training load, mood depression, and frequent infections

Table 8.2 Commonly reported symptoms of overtraining.

Underperformance	Increased early morning or sleeping heart rate
Muscle weakness	Altered mood (greater depression, anger, confusion)
Chronic fatigue	Loss of appetite
Sore muscles	Gastrointestinal disturbance
Recurrent infection	Increased perceived exertion during exercise
Reduced motivation	Sleep disturbance

The consequences of overtraining range from altered muscle function to motivation. The pathophysiology of overtraining can include muscle soreness and weakness, cytokine actions, hormonal and haematological changes, mood swings and psychological depression, and nutritional problems such as loss of appetite and diarrhoea. Many symptoms have been reported in overtrained athletes; the more commonly reported symptoms are listed in Table 8.2. In some cases the underlying cause could be a persistent viral infection, similar to glandular fever, or a type of post-viral fatigue syndrome not unlike myalgic encephalomyelitis (ME). Athletes suffering from overtraining syndrome are often reported to be immunodepressed.

The temporary underperformance associated with overreaching may be the result of insufficient recovery of muscle glycogen and/or exercise-induced muscle damage; many athletes experiencing overreaching (and some who have been diagnosed as suffering from overtraining) report that their muscles feel sore. The consequences of exercise-induced muscle damage include delayed-onset muscle soreness and stiffness, reduced range of motion, higher than normal blood lactate concentration and perceived exertion during exercise, loss of strength, and reduced maximal dynamic power output that can last 5–10 days. A practical index of muscle fibre damage is an elevation of muscle proteins [e.g. myoglobin, creatine kinase (CK), lactate dehydrogenase, and myosin heavy-chain fragments] in the blood plasma. However, well-trained athletes who perform eccentric muscle actions do not usually show large increases in plasma CK, although they still experience soreness, perhaps as a result of damage and inflammation of connective tissue structures in muscle.

The damaged muscle tissue can cause an initial activation of the immune system, as white blood cells are attracted to the damaged muscles to begin breakdown of damaged fibres and initiate the repair process. Continuing to undertake strenuous exercise with already damaged muscles may make the situation even worse by preventing tissue repair and maintaining a state of inflammation. Another practical problem is that 3–6 days after a muscle-damaging bout of exercise, athletes can no longer perceive that their muscles are still weak, so they may exercise too hard too soon. Weakness also makes the muscle more prone to injury.

Another detrimental effect of exercise-induced muscle damage is that it impairs the restoration of muscle glycogen (a polymer of glucose that serves as an important energy source for exercise). Stores of glycogen become depleted after prolonged exercise. Damaged muscle has an impaired ability to take up glucose from the blood which is required to resynthesize glycogen in the muscle. This would be expected to result in decreased endurance performance in subsequent exercise bouts.

While there is a place for eccentric exercise, plyometrics, and weightlifting in the training programmes of many athletes, the frequency and intensity of these exercise modes should be carefully regulated and positively avoided in the weeks leading up to competition. Some individuals are far more prone to muscle soreness and stiffness than others, although there is no evidence to link this to any form of rheumatism or arthritis. A thorough warm-up and slow-stretching techniques applied both before and immediately after exercise can minimize subsequent stiffness.

There is no single or reliable marker of impending overtraining

Reliable techniques for the detection of the onset of overtraining have not yet been established. Possible markers are being studied, including blood levels of stress hormones, antibodies, cytokines, and amino acids as well as the ability of white blood cells to respond to stimulation by microorganisms. Sleeping heart rate and psychological profiling are other approaches that have been tried and may be of some use. Measures of these potential markers made in athletes undertaking their normal training and in others whose training loads have been markedly increased, as well as in athletes who are diagnosed to be currently suffering from overtraining syndrome, may enable sports scientists to screen athletes for the onset of overtraining. While no single marker can be taken as an indicator of impending overtraining, the regular monitoring of a combination of performance, physiological, biochemical, immunological, and psychological variables would seem to be the best strategy to identify athletes who are failing to cope with the stresses of training. Again it is important to stress the need to distinguish overtraining from overreaching and other potential causes of temporary underperformance such as anaemia, acute infection, and insufficient carbohydrate intake.

Exercise training, immune function, and susceptibility to infection

Endurance athletes engaged in heavy training programmes appear to be more susceptible than normal to infection

Athletes engaged in heavy training programmes, particularly those involved in endurance events, appear to be more susceptible than normal to infection. For example, several surveys have described a substantially higher (2–6-fold) frequency of self-reported symptoms of upper respiratory tract infection (URTI) in athletes who completed long-distance foot races compared with control runners who did not compete in the events. There is some convincing evidence that this increased susceptibility to infection arises due to a depression of immune system function (for detailed reviews see Gleeson and Bishop 1999; Gleeson 2007).

Acute effects of exercise on immune function

Prolonged strenuous exercise depresses several aspects of immune cell function for several hours

Prolonged strenuous exercise has a temporary depressive effect on immune function and this has been associated with an increased incidence of infection. An acute bout of physical activity is accompanied by responses that are remarkably similar in many respects to those induced by infection, sepsis, or trauma: there is a substantial increase in the number of circulating leukocytes (mainly lymphocytes and neutrophils), the magnitude of which is related to both the intensity and duration of exercise. There are also increases in the plasma concentrations of various substances that are known to influence leukocyte functions, including inflammatory and anti-inflammatory cytokines such as tumour necrosis factor-α, interleukin (IL)-1β, IL-6, IL-10, macrophage inhibitory protein-1 and IL-1-receptor antagonist, acute phase proteins such as C-reactive protein, and activated complement fragments. The large increases in plasma IL-6 concentration observed during exercise can be entirely accounted for by release of this cytokine from activated muscle fibres. However, IL-6 production by monocytes and IL-2 and interferon-gamma (IFN-γ) (but not IL-4) production by T lymphocytes is inhibited during and for several hours after prolonged exercise. Hormonal changes also occur in response to exercise, including rises in the plasma concentration of several hormones (e.g. adrenaline, cortisol, growth hormone, and prolactin) which are known to have immunomodulatory effects. Phagocytic neutrophils appear to be activated by an acute bout of exercise but show a diminished responsiveness to stimulation by bacterial lipopolysaccharide and reduced oxidative burst (killing capacity) after exercise, which can last for many hours. Acute exercise temporarily increases the number of circulating natural killer (NK) cells, but following exercise NK cell counts drop to less than half of normal levels for a couple of hours; normal resting values are usually restored within 24 h. NK cell cytolytic activity (per cell) falls after exercise, and if the activity is both prolonged and vigorous, the decrease in NK cell counts and cytolytic activity may begin during the exercise session. During recovery from exercise, lymphokine-activated killer (LAK) cell numbers and activity also fall below pre-exercise levels. Acute exercise has been shown to diminish the proliferative response of lymphocytes to mitogens and decrease the expression of an early activation marker (CD69) in response to stimulation with mitogen. When the exercise bout is strenuous and very prolonged (>1.5 h), the number of circulating lymphocytes may be decreased below pre-exercise levels for several hours after exercise and the T-lymphocyte CD4+/CD8+ ratio is decreased. Both T memory (CD45RO+) and T naive (CD45RA+) cells increase temporarily during exercise, but the CD45RO/CD45RA ratio tends to increase due to the relatively greater mobilization of the CD45RO+ subset. Following prolonged strenuous exercise the production of immunoglobulins by B lymphocytes is inhibited and skin inflammatory response to antigens is diminished. After prolonged exercise, the plasma concentration of glutamine has been reported to fall by about 20% and may remain depressed for some time. These changes during early recovery from exercise would appear to weaken the potential immune response to pathogens and have been suggested to provide an 'open window' for infection, representing the most vulnerable time period for an athlete in terms of their susceptibility to contracting an infection.

Chronic effects of exercise training on immune function

Periods of intensified training are associated with the development of chronic immunodepression

Exercise training also modifies immune function, with most changes on balance suggesting an overall decrease in immune system function, particularly when training loads are high. Circulating numbers of leukocytes are generally lower in athletes at rest compared with sedentary people, although there is a weak suggestion of slightly elevated NK cell counts and cytolytic action in trained individuals. A low blood leukocyte count may arise from the haemodilution (expansion of the plasma volume) associated with training, or may represent altered leukocyte kinetics including a diminished release from the bone marrow. Indeed, the large increase in circulating neutrophil numbers that accompanies a bout of prolonged exercise could, over periods of months or years of heavy training, deplete the bone marrow reserve of these important cells. Certainly, the blood population of these cells seems to be less mature than those found in sedentary individuals, and the phagocytic and oxidative burst activity of stimulated neutrophils has been reported to be markedly lower in well-trained cyclists compared with age- and weight-matched sedentary controls. Levels of secretory immunoglobulins such as salivary immunoglobulin A (IgA) are lower in athletes engaged in heavy training, as are T-lymphocyte helper/suppressor (CD4+/CD8+) ratios and *in vitro* mitogen-stimulated lymphocyte proliferation responses, although exercise training in healthy young adults does not appear to have an effect on the initiation of a specific antibody response to vaccination or skin inflammatory responses to antigens.

Thus, with chronic periods of heavy training, several aspects of both innate and adaptive immunity are depressed.

> Elevated levels of stress hormones cause a temporary inhibition of IFN-γ production by T lymphocytes; this may be mostly responsible for exercise-induced depression of immune cell functions

There are several possible causes of the diminution of immune function associated with heavy training. One mechanism may simply be the cumulative effects of repeated bouts of intense exercise with the consequent elevation of stress hormones, particularly glucocorticoids, such as cortisol, causing temporary immunosuppression. It is known that both acute glucocorticosteroid administration and exercise cause a temporary inhibition of IFN-γ production by T lymphocytes and it has been suggested that this may be an important mechanism in exercise-induced depression of immune cell functions. When exercise is repeated frequently there may not be sufficient time for the immune system to recover fully. Furthermore, plasma glutamine levels can change substantially after exercise and may become chronically depressed after repeated bouts of prolonged strenuous training. Complement activation also occurs during exercise and a diminution of the serum complement concentration with repeated bouts of exercise, particularly when muscle damage is incurred, could also contribute to decreased non-specific immunity in athletes; well-trained individuals have a lower serum complement concentration compared with sedentary controls.

In summary, acute bouts of exercise cause a temporary depression of various aspects of immune function [e.g. neutrophil respiratory burst, lymphocyte proliferation, monocyte major histocompatibility complex class II (MHCII) expression] that lasts approximately 3–24 h depending on the intensity and duration of the exercise bout. Periods of intensified training (overreaching) lasting 7 days or more result in chronically depressed immune function.

Effects of detraining

> Training adaptations are reversible

When training stops, either voluntarily or because of injury, there is a prompt reversal of the adaptations that occurred during the training period. There are, however, some important differences, especially in the time course of the changes that take place. You can see an illustration of this in Figure 8.9, which shows that whole-body aerobic capacity is reasonably well maintained for some time after training stops, but some variables can be lost within an alarmingly short time. For example, about 50% of the increased mitochondrial content induced by

Figure 8.9

Aerobic capacity is well maintained for some time after training stops but some of the muscle adaptations are lost fairly quickly.

training can be lost within 1–2 weeks of detraining. Even a few short periods of high-intensity exercise seem to be effective in preventing—or at least slowing— the decline in function that accompanies detraining. A return to training will, of course, allow recovery of the muscle adaptations, but the time required to re-establish the original trained condition can take somewhat longer than the detraining interval over which they were lost.

Nutritional effects on training adaptation

Training and diet

Dietary strategies can influence the effectiveness of a training programme by promoting metabolic adaptations

It does not seem sensible to train hard to improve performance and to neglect the gains that can be made by attention to diet. Although genetic endowment and training are the key factors in improving performance, diet has a key role. Diet can support the training load while minimizing the risk of chronic fatigue, illness, and injury. It is now being increasingly recognized that dietary strategies can influence the effectiveness of a training programme by promoting the metabolic adaptations that take place in the trained tissues.

Nutrition needs in training are influenced by several different factors, including the intensity, duration, and frequency of the training bouts, and by the objectives of the training programme. Some of the key issues have been covered in the preceding chapters, and it is noticeable that there are more similarities than differences in the nutritional needs of athletes training for different events.

Nutrient–gene interactions

Exercise upregulates the gene expression of a number of genes that encode for proteins in skeletal muscle that play a role in meeting the demands of exercise

Exercise physiologists have been interested in the ability of skeletal muscle to adapt to repeated bouts of exercise for many decades. The attempts to understand the basic mechanisms that regulate these changes have increased tremendously in the past 10 years with the increasing use of powerful molecular and cellular analytical techniques that allow the quantification of specific gene expression. Numerous investigations have reported that exercise upregulates the gene expression of a number of genes that encode for proteins in skeletal muscle which play a role in meeting the demands of exercise (see Hargreaves and Cameron-Smith 2002 for a detailed review). It has been suggested that the cellular adaptations to exercise training may be due to the cumulative effects of these transient increases in gene transcription that occur during and after repeated exercise bouts. In this section of the chapter we are particularly interested in the link between nutrition and the adaptations that occur in human skeletal muscle during training. Some recent examples of the consequences of short-term dietary manipulations on gene expression in human skeletal muscle are discussed in the following section.

Low levels of muscle glycogen and high-fat diets may enhance the mRNA content of some genes involved in exercise metabolism

Peters *et al.* (2001) examined the effects of consuming an isoenergetic high-fat diet (73% fat, 5% CHO, 22% protein) following a standardized 'normal' diet (30% fat, 50% CHO, 21% protein) on mRNA specific for pyruvate dehydrogenase kinase (PDK), PDK protein content, and PDK activity in human skeletal muscle. PDK is an enzyme that inhibits the activity of PDH, the enzyme that converts pyruvate into acetyl-CoA. Thus, an increase in PDK activity will reduce the rate of CHO oxidation. Muscle biopsies were taken from the vastus lateralis muscle of six subjects before the high-fat diet (day 0) and each morning for up to 3 days on the high-fat diet. PDK activity increased after only 1 day on the high-fat diet and continued to increase on successive days on the high-fat diet. Both mRNA and protein PDK levels increased dramatically following 1 day on the high-fat diet and remained high with no further increases on days 2 or 3. This is an impressive adaptive quality for an enzyme that plays such an important role in the oxidation of CHO both at rest and during exercise. Muscle appears to quickly respond to a lack of CHO availability and/or increased fat oxidation by upregulating PDK activity, which drives more of the PDH enzyme into the inactive form and ultimately decreases skeletal muscle CHO oxidation and spares the small store of CHO in the body. Experiments that have artificially elevated the plasma FFA levels for a 4–5-h period have reported very large increases in PDK mRNA levels, emphasizing how rapidly fuel availability can upregulate PDK gene expression and decrease CHO oxidation in skeletal muscle.

Cameron-Smith *et al.* (2003) examined the effects of either a high-CHO diet (70–75% CHO, <15% fat) or an isoenergetic high-fat diet (>65% fat, <29% CHO) for 5 days on the expression of genes encoding proteins for fatty acid transport and β-oxidation of FFA in human skeletal muscle. The subjects were 14 well-trained cyclists and triathletes who continued to train daily during the 5-day diet interventions. Plasma FFA levels were higher following 5 days on the high-fat diet compared with baseline and following the high-CHO diet. CHO oxidation was reduced and fat oxidation was increased during cycling at about 70% VO_{2max} following the high-fat diet compared with the high-CHO diet. There were significant increases in the fatty acid transporter (FAT/CD36) and β-hydroxyacyl-CoA dehydrogenase mRNA levels following the high-fat diet as compared to the high-CHO and baseline conditions. FAT/CD36 protein content was also increased in muscles from eight subjects following the high-fat diet. These data provide strong support for the idea that increased dietary fat intake can increase the mRNA levels of genes that are necessary for the uptake and oxidation of FFA. This muscle adaptation seems entirely appropriate in the face of reduced CHO availability and increased plasma FFA availability. It is also noteworthy that these changes occurred in very well-trained athletes, who presumably have maximized the ability of their muscles to oxidize fat through years of endurance training. However, some caution is necessary in interpreting the results of these studies as increases in mRNA levels are not always predictive of increases in protein or measures of functional activity. There are many additional steps and points of regulation between increased mRNA contents and increased protein synthesis rates.

Low levels of muscle glycogen may enhance the mRNA content of some genes involved in exercise metabolism. A recent study manipulated pre-exercise muscle glycogen levels with a combination of exercise and diet and found that the genes for selected metabolic enzymes including the PDK gene were upregulated to a greater extent in response to exercise when exercise was performed with low pre-exercise glycogen content. The role of muscle glycogen could be explained by the fact that some signalling proteins (e.g. AMPK) possess glycogen binding domains, and when glycogen is low, these proteins are more active towards their specific targets. Indeed, the activity of both AMPK and p38 mitogen-activated protein kinase (p38 MAPK) is elevated when exercise is performed with low muscle glycogen. This may be beneficial to individuals undertaking endurance exercise training as AMPK and p38 MAPK are believed to play a critical role in regulating mitochondrial biogenesis and endurance training adaptations. However, it is not yet known whether these exaggerated responses in a glycogen-depleted state translate into greater protein contents or higher functional activities. It is likely that signals responsive both to increased fat availability and to decreased CHO availability would work in concert to determine the exact responses in gene expression in skeletal muscle. The mechanisms by which the adaptations to a high-fat/low-CHO diet are mediated appear to be related to FFA activation of the family of peroxisome proliferator activator receptors (PPARs) and/or an insulin effect consequent to the decreased CHO availability.

In summary, it seems clear that nutrition can alter gene expression both at rest and in combination with exercise. However, it is important to note that the dietary manipulations discussed above were for the most part drastic and unlikely to be used by athletes actively engaging in training and competition. It remains to be seen whether smaller changes in the dietary CHO and fat content will have effects on gene expression independent of or in combination with the training-induced changes.

Train low, compete high: endurance training in a glycogen-depleted state results in an improved capacity to use fat to fuel exercise

To date, there is limited evidence to determine whether long-term training with low muscle glycogen can enhance the adaptive response to endurance training. In one study, untrained men undertook a 10-week programme of knee-extensor exercise. Each of the subjects' legs was trained according to a different schedule but the total amount of work performed by each leg was kept the same. One leg was trained twice-a-day every second day, whereas the other leg trained once daily. This meant that the leg that trained twice daily commenced half of the sessions with low muscle glycogen. Compared with the leg that trained with normal glycogen levels, the leg that commenced half of the training sessions with low muscle glycogen had more pronounced increases in resting muscle glycogen content and citrate synthase activity. Time to fatigue at 90% maximal power output increased in both legs after training. However, performance times were nearly twice as long in the leg that trained with low muscle glycogen. Thus, under the specific conditions of the study, training with low muscle glycogen enhanced adaptations in skeletal muscle and improved exercise performance. However, it should be borne in mind that the subjects recruited were untrained and it is not yet known if training with low muscle glycogen will translate into improved adaptations in already well-trained athletes. Secondly, subjects performed a fixed amount of work even though higher glycogen stores would normally allow for exercise to be performed at higher intensities and/or longer durations. Thirdly, it is difficult to translate the results from single-leg kicking exercise to that of real-life sporting situations involving activities such as running, cycling, or swimming.

A more recent study has confirmed that when 3 weeks' of bicycle training were performed with all the training done in the glycogen-loaded state or where half of the training was performed in a glycogen-depleted state, the performance during the self-paced training sessions in the glycogen-depleted state was significantly impaired. The subjects who trained always glycogen-loaded trained at higher absolute intensities. However, despite this, at the end of the training period performance was equally improved in the low glycogen and high glycogen groups. The metabolic adaptations between the two groups were very different though. The low glycogen group had a greater oxidative capacity as evidenced by greater citrate synthase and β-hydroxyacyl-CoA dehydrogenase. Fat oxidation during exercise was also enhanced in the low glycogen group. So these findings partly confirm the findings of the single leg study by Hansen *et al.* The difference is that even though adaptations were observed in metabolism, there were no differences in performance after three weeks. It is of course possible that three weeks is sufficient to see differences in metabolism but not quite long enough to see improvements in performance and future studies need to examine longer-term training regimens.

So, it seems fair to conclude that endurance training in a glycogen-depleted state results in an improved capacity to use fat to fuel exercise. The mechanisms behind this are unclear but may involve activation of the PPARs—the group of nuclear receptor proteins that function as transcription factors regulating the expression of genes and play an essential role in the regulation of metabolism as well as many other processes. Training rats on a treadmill while at the same time giving them a drug that activated a transcription factor called PPAR∂ has been shown to result in the same changes that occur when training in the glycogen-depleted state: increased capacity to use fat as a fuel. PPAR∂ increases the concentration of the enzymes that are required for oxidizing fatty acids. The result in this study was that the rats that both got the drug and trained on the treadmill increased their ability to run at ~50% VO_{2max} by 70% over those that just ran on the treadmill. It is possible that exercising in a glycogen-depleted state activates PPAR∂ to a greater extent than training in the glycogen-loaded state. PPAR∂ seems to be activated by a byproduct of the breakdown of fat in muscle. As discussed above, exercising in the glycogen-depleted state increases circulating FA concentration and the oxidation of fat during exercise resulting in more of the byproduct and more PPAR∂ activation.

Although the results of some recent studies suggest that training with low glycogen could have some metabolic advantages, this does not seem to result in improvements in performance. There is a possible disadvantage to training with low glycogen in that stress hormone responses to exercise will be elevated compared with exercise in a glycogen-loaded state, which is likely to be associated with greater immune pertubations and earlier development of overreaching symptoms.

Free radicals and antioxidants

Free radicals are highly reactive and potentially damaging molecules

A free radical is a molecule that has an unpaired electron (represented by a dot next to the chemical structure, e.g. **A**·) as illustrated in Figure 8.10. Some radicals

Figure 8.10

Free radicals have an unpaired electron in their outer orbit and this is what makes them highly reactive.

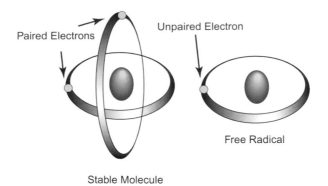

Paired Electrons

Unpaired Electron

Free Radical

Stable Molecule

are stable and long-lived (the most common example is O_2, the oxygen in the air we breathe, which is a 'biradical' as it has two unpaired electrons), but most radicals are highly reactive and thus unstable. As a rule, a radical needs to pair its unpaired electron with another, and will react with another molecule in order to obtain this missing electron. If a radical achieves this by 'stealing' an electron from another molecule, that other molecule itself becomes a radical (Reaction 1), and a self-propagating chain reaction is begun (Reaction 2).

$$A^{\cdot} + B{:} \rightarrow A{:} + B^{\cdot} \quad \textbf{(Reaction 1)}$$

$$B^{\cdot} + C{:} \rightarrow B{:} + C^{\cdot} \quad \textbf{(Reaction 2)}$$

If a radical pairs its unpaired electron by reacting with a second radical, then the chain reaction is terminated, and both radicals 'neutralize' each other (Reaction 3).

$$A^{\cdot} + B^{\cdot} \rightarrow A{:}B \quad \textbf{(Reaction 3)}$$

Radicals are produced by normal aerobic metabolism and are necessary to life. However, excess amounts of radicals are harmful because of their reactivity. Radicals can be formed easily by compounds that readily give up a single electron (for example, polyunsaturated fatty acids). Radicals are also produced by processes other than normal metabolism; for example, by ionizing radiation, smoking and other pollutants, herbicides and pesticides, and are even found in certain types of food (e.g. deep-fat fried foods). Radicals can damage lipids (fats, which make up cell membranes), proteins, and DNA. Radical damage can be significant because it can proceed as a chain reaction.

Oxygen in the air we breathe is in its 'ground' (not energetically excited) state and is symbolized by the abbreviation $^3O_2$. It is a free radical—in fact it is a diradical, as it has two unpaired electrons. Electrons 'spin' (that is, rotate about an axis passing through the electron). Molecules whose outermost pair of electrons have parallel spins (symbolized by ↑↑) are in the 'triplet' state; molecules whose outermost pair of electrons have antiparallel spins (symbolized by ↑↓) are in the 'singlet' state. Ground-state oxygen is in the triplet state (indicated by the superscripted '3' in $^3O_2$)—its two unpaired electrons have parallel spins, a characteristic that does not allow them to react with most molecules. Thus, ground-state or triplet oxygen is not very reactive. However, triplet oxygen can be activated by the addition of energy, and transformed into reactive oxygen species. If triplet oxygen absorbs sufficient energy to reverse the spin of one of its unpaired electrons, it forms the singlet state. Singlet oxygen, abbreviated $^1O_2^*$, has a pair of electrons with opposite spins; though not a free radical it is highly reactive. The * symbol is used to indicate that this is an excited state with excess energy. It should be emphasized that neither triplet- nor singlet-state molecules are necessarily in the excited state; the designation of 'triplet' or 'singlet' refers only to the spin state.

$$\cdot O\text{-}O\cdot \text{ (triplet oxygen)} \rightarrow O\text{-}O: \text{ (singlet oxygen)}$$

ground state highly reactive

This reaction can also be written in this form:

$$^3O_2 + \text{energy} \rightarrow {}^1O_2^*$$

Singlet oxygen is produced as a result of natural biological reactions and by photosensitization by the absorption of light energy. According to rules of physical chemistry, the 'relaxation' (excess energy loss) of singlet oxygen back to the triplet state is 'spin forbidden' and thus singlet oxygen has a long lifetime for an energetically excited molecule, and must transfer its excess energy to another molecule in order to relax to the triplet state.

Triplet oxygen can also be transformed into a reactive state if it accepts a single electron. This process of accepting an electron is called **reduction**, and in this case, is 'monovalent' reduction because only one electron is involved. The molecule that gave up the electron is **oxidized**. The result of monovalent reduction of triplet oxygen is called a **superoxide**, abbreviated O_2^-. Superoxide is a radical. It is usually shown with a negative sign, indicating that it carries a negative charge of –1 (due to the extra electron, e^-, it gained).

$$\cdot O-O\cdot \text{ (triplet oxygen)} \rightarrow \cdot O-O: \text{ (superoxide)}$$

This reaction can also be written in this form:

$$^3O_2 + e^- \rightarrow O_2^{\cdot -}$$

Superoxide can act both as an oxidant (by accepting electrons) or as a reductant (by donating electrons). However, superoxide is not particularly reactive in biological systems and does not by itself cause much oxidative damage. It is a precursor to other oxidizing agents, including singlet oxygen, peroxynitrite, and other highly reactive molecules. However, superoxide is not all bad—in fact it is necessary for health. For example, certain cells in the human body produce superoxide (and the reactive molecules derived from it) as a chemical weapon used to kill invading microorganisms. Superoxide also acts as a signalling molecule needed to regulate some cellular processes. Under biological conditions, the main reaction of superoxide is to react with itself to produce hydrogen peroxide and oxygen, a reaction known as 'dismutation'. Superoxide dismutation can be spontaneous or can be catalysed by the enzyme superoxide dismutase ('SOD'):

$$O_2^{\cdot -} + 2H^+ \rightarrow H_2O_2 + {}^3O_2$$

Superoxide is also important in the production of the highly reactive hydroxyl radical ($\cdot$OH) (discussed in more detail below). In this process, superoxide actually acts as a reducing agent, not as an oxidizing agent. This is because superoxide donates one electron to reduce the metal ions (ferric iron or Fe^{3+} in the example below) that act as the catalyst to convert hydrogen peroxide (H_2O_2) into the hydroxyl radical ($\cdot$OH):

$$O_2^{\cdot -} + Fe^{3+} \rightarrow {}^3O_2 + Fe^{2+}$$

The reduced metal (ferrous iron or Fe^{2+} in this example) then catalyses the breaking of the oxygen–oxygen bond of hydrogen peroxide to produce a hydroxyl radical ($\cdot$OH) and a hydroxide ion (OH^-):

$$Fe^{2+} + H_2O_2 \rightarrow Fe^{3+} + {}^\cdot OH + OH^-$$

Superoxide can react with the hydroxyl radical ($\cdot$OH) to form singlet oxygen ($^1O_2^*$) (not a radical but reactive nonetheless):

$$O_2^{\cdot -} + {}^\cdot OH \rightarrow {}^1O_2^* + OH^-$$

Superoxide can also react with nitric oxide ($NO^\cdot$) (also a radical) to produce peroxynitrate ($OONO^-$), another highly reactive oxidizing molecule:

$$O_2^{\cdot -} + NO^\cdot \rightarrow OONO^-$$

Superoxide ($O_2^{\cdot -}$) can undergo monovalent reduction to produce **peroxide** (O_2^{-2}), an activated form of oxygen that carries a negative charge of −2. Usually peroxide is termed 'hydrogen peroxide' (H_2O_2) since in biological systems the negative charge of −2 is neutralized by two protons (two hydrogen atoms, each with a positive charge):

$$^\cdot O\text{-}O\text{:} \rightarrow H\text{:}O\text{-}O\text{:}H$$

Hydrogen peroxide is important in biological systems because it can pass readily through cell membranes and cannot be excluded from cells. Hydrogen peroxide is actually necessary for the function of many enzymes, and thus is required (like oxygen itself) for health. Hydrogen peroxide is not as reactive as a product it can form, the hydroxyl radical. Hydrogen peroxide, in the presence of metal ions, is converted to a hydroxyl radical ($^{\bullet}OH$) and a hydroxide ion (OH^-). The metal ion is required for the breaking of the oxygen–oxygen bond of peroxide. This reaction is called the Fenton reaction, and was discovered over a hundred years ago. It is important in biological systems because most cells have some level of iron, copper, or other metals that can catalyse this reaction. This reaction can also be written (with iron as the metal):

$$Fe^{2+} + H_2O_2 \rightarrow Fe^{3+} + {}^{\bullet}OH + OH^- \text{ (the Fenton reaction)}$$

A hydroxyl radical can also react with superoxide to produce singlet oxygen and a hydroxide ion:

$$O_2^{\bullet -} + {}^{\bullet}OH \rightarrow {}^1O_2^* + OH^-$$

Like hydrogen peroxide, the hydroxyl radical passes easily through membranes and cannot be kept out of cells. Hydroxyl radical damage is 'diffusion rate-limited'. Simply put, the rate at which hydroxyl radicals can damage other molecules is limited chiefly by how far and fast it can diffuse (travel) in the cell. This highly reactive radical can add to an organic (carbon-containing) substrate (represented by **R** below)—this could be for example a fatty acid forming a hydroxylated adduct that is itself a radical:

$${}^{\bullet}OH + R \rightarrow HOR^{\bullet}$$

The hydroxyl radical can also oxidize the organic substrate by 'stealing' or abstracting an electron from it:

$${}^{\bullet}OH + R \rightarrow R^{\bullet} + H_2O$$

The resulting oxidized substrate is again itself a radical, and can react with other molecules in a chain reaction. For example, it could react with ground-state oxygen to produce a peroxyl radical ($ROO^{\bullet}$):

$$R^{\bullet} + {}^3O_2 \rightarrow ROO^{\bullet}$$

The peroxyl radical again is highly reactive, and can react with another organic substrate in a chain reaction:

$$ROO^{\bullet} + RH \rightarrow ROOH + R^{\bullet}$$

This type of chain reaction is common in the oxidative damage of fatty acids and other lipids, and demonstrates why radicals such as the hydroxyl radical can cause so much more damage than one might have expected. Similar damage caused by hydroxyl radicals and other reactive oxygen species can occur in proteins and with nucleic acids (mainly DNA). Proteins are highly susceptible to oxidative damage, particularly at sites where sulphur-containing amino acids are found. DNA can be oxidatively damaged at both the nucleic bases (the individual molecules that make up the genetic code) and at the sugars that link the bases. Oxidative damage of DNA results in degradation of the bases, breaking of the DNA strands by oxidation of the sugar linkages, or cross-linking of DNA to protein (a form of damage particularly difficult for the cell to repair). Although all cells have some capability of repairing oxidative damage to proteins and DNA, excess damage can cause mutations or cell death.

> The body tissues continually produce reactive oxygen species and other free radicals

In normal cellular metabolism, small amounts of ROS are produced during the aerobic process by which humans and animals derive energy from the mitochondria. In the electron-transport chain, located on the inner mitochondrial membrane, most of the oxygen consumed by cells is reduced via cytochrome oxidase to yield water and energy that is used to resynthesize ATP from ADP and P_i. However, a small proportion, most recently estimated to be about 0.15% of the total oxygen consumed at rest can be used in an alternative pathway for the univalent reduction of oxygen, and thus, the superoxide radical ($O_2^{\cdot-}$) is produced (Figure 8.11). Remember that one free-radical molecule can start a destructive process by removing electrons from stable compounds, such as polyunsaturated FFAs, and forming large numbers of ROS, thereby transforming stable compounds into highly reactive free radicals.

Additional sources of ROS and other free radicals are ultraviolet light, alcohol, cigarette smoke, high-fat diets, and the respiratory burst of white blood cells, such as neutrophils and monocytes, that are capable of ingesting foreign material, including bacteria. The generation of ROS by the neutrophil respiratory burst is an essential mechanism of the host defence mechanism of the immune system for killing bacteria and clearing away damaged tissue. However, these ROS can also initiate damaging chain reactions, such as lipid peroxidation, and the subsequent release of large numbers of ROS. In addition during whole-body eccentric exercise, invading neutrophils and monocytes are thought to be important sources of oxidative stress. Migration of neutrophiles into damaged muscle tissue occurs within the first hour after eccentric exercise and is followed by infiltration of monocytes and macrophages, reaching a maximum at 24 h to 72 h after exercise. Several inflammatory diseases have also been associated with increased generation of ROS by monocytes.

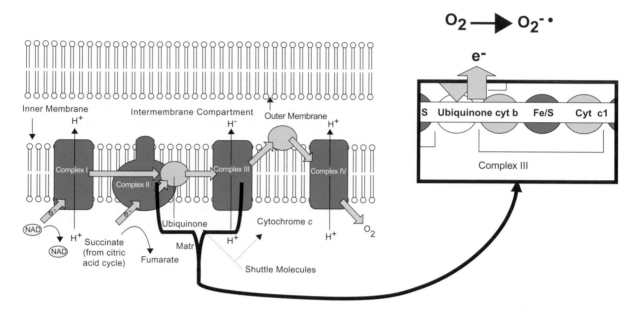

Figure 8.11

Formation of superoxide in the mitochondrial electron-transport chain.

Increases in the free iron concentration of body fluids (caused, for example, by haemolysis or ischaemia-reperfusion) may amplify ROS toxicity by increasing the generation of the highly reactive hydroxyl radical (˙OH). Ferric iron (Fe^{3+}) ions can stimulate free-radical reactions by breaking down lipid peroxides to chain-breaking alkoxyl radicals, and ferrous (Fe^{2+}) ions can react with H_2O_2 to give ˙OH and other highly reactive species in the previously mentioned Fenton reaction. The hydroxyl radical can also be formed through the Haber–Weiss reaction:

$$Fe^{2+} + H_2O_2 + O_2^{˙-} \rightarrow Fe^{3+} + \text{˙OH} + OH^- + O_2$$

This is one of the reasons why iron supplements should not be recommended indiscriminately. Free radicals have been implicated in the aetiology of various diseases including cancer and coronary heart disease and an excess of iron could potentiate their adverse effects.

> Free radicals are produced in increased amounts during exercise

The first serious studies of the generation and potential roles of ROS during exercise were carried out in the late 1970s and during the ensuing years, a great deal of research has been undertaken to try and understand the nature and sources of the species generated, the factors influencing their generation, their effects on muscle and other cells, and how these effects might be manipulated. An assumption that has underpinned much of the work since the early studies is that the species generated are essentially byproducts of metabolism and damaging to cells and tissues, and it is notable that even the earliest studies in the area attempted to scavenge the species generated and look for potential functional benefits of such interventions. Unsurprisingly, early studies were also characterized by limitations in the analytical methods available to detect free-radical species, and non-specific approaches such as the analysis of lipid peroxides such as malondialdehyde or thiobarbituric acid reactive species (TBARS) in complex biological tissues were commonly used. These studies generally assumed that the predominant source for formation of free radicals during exercise is from leakage from the electron-transport chain in the mitochondria (as much as 3–5% of the total oxygen consumed by mitochondria was suggested to undergo one electron reduction with the generation of superoxide). It was also assumed that the amount of ROS generated in this way was directly related to the rate of oxygen uptake by muscle tissue that can increase up to 100-fold during exercise compared with rest. This, of course implies that potentially a 100-fold increase in superoxide generation by skeletal muscle could occur during aerobic exercise. These assumptions became firmly rooted and extensively quoted in the subsequent literature, but have not been supported by the most recent research in this area.

Since 2000, the development of improved techniques to assess the primary ROS generated in skeletal muscle has established that the primary ROS generated by skeletal muscle are nitric oxide and superoxide. Mitochondria are still thought to be the major site of superoxide generation in tissues but a number of recent findings have argued against the previously high estimates of rates of formation of superoxide within mitochondria. The most recent estimates of the rate of production of ROS by mitochondria indicate that no more than about 0.15% of the electron flow gives rise to ROS (i.e. less than 10% of the original minimum estimate), and it is becoming increasingly clear that even this low rate of production may be further reduced by intrinsic control mechanisms. As for exercise, the most recent data indicate that muscle intracellular ROS production increases by only a modest 2- to 4-fold during contractions, which seems to support the suggestion that there is considerable internal control of mitochondrial ROS generation.

Studies have identified the NAD(P)H oxidase enzymes associated with the sarcoplasmic reticulum (SR) of cardiac and skeletal muscle as additional sources of ROS. The superoxide generated by these enzymes seems to influence calcium release by the SR through oxidation of the ryanodyne receptor. A NAD(P)H oxidase somewhat similar to that found in phagocytic cells of the immune system has also been described located in the plasma membrane, triads, and transverse tubules of skeletal muscle that is activated by membrane depolarization. Many studies have now indicated that skeletal muscle cells release superoxide into

Figure 8.12

Simplified diagram of the processes contributing to free-radical generation in exercising muscle. ETC: electron transport chain.

the extracellular space and non-muscle cells contain other plasma membrane redox systems capable of undertaking electron transfer across the plasma membrane to effect transfer of electrons from intracellular reductants to appropriate extracellular electron acceptors. Other enzyme systems that can generate ROS in skeletal muscle tissue include phospholipases and xanthine oxidase. Skeletal muscle fibres themselves have been reported not to contain significant amounts of xanthine oxidase, although this enzyme will inevitably be present in associated endothelial cells of the blood vessels in muscle tissue. The various potential pathways of ROS generation within muscle are illustrated in Figure 8.12.

Various antioxidant mechanisms protect tissues against an excess of free radicals

An antioxidant is a compound that protects biological systems against the harmful effects or reactions that create excessive oxidants. Dietary antioxidants significantly decrease the adverse effects of ROS. Enzymatic and non-enzymatic antioxidants prevent oxidation initiated by ROS by:

- preventing ROS formation;
- intercepting ROS attack by scavenging the reactive metabolites and converting them to less-reactive molecules;

- binding transition metal ion catalysts, such as copper and iron, to prevent initiation of free-radical reactions;
- reacting with chain-propagating radicals, such as the peroxyl and alkoyl species, to prevent continued hydrogen abstraction from FA side chains; and
- providing a favorable environment for the effective functioning of other antioxidants or acting to regenerate the nonenzymatic antioxidant molecules.

Antioxidant enzymes can affect free-radical generation at both the initiation and propagation stage. The antioxidant enzymes superoxide dismutase (SOD) and catalase can inhibit the initial phase by inactivating precursor molecules of ROS production (see Figure 8.13). At the propagation stage, glutathione peroxidase—another antioxidant enzyme—can scavenge ˙OH and lipid peroxides

Figure 8.13

The antioxidant action cascade.

as described earlier. The trace elements copper and manganese are required as cofactors of SOD, and selenium is also a component of antioxidant defence because it is a cofactor of glutathione peroxidase/reductase and, thus, influences the quenching of ROS. The effects of exercise on manganese and selenium status are presently unknown, but training is associated with an increase in levels of antioxidant enzymes, suggesting that there may be an increased requirement for these trace elements during periods of increased training.

In addition to the enzyme defence system, some low molecular weight substances act as scavengers of radicals. These substances include vitamins A, C, and E; carotenoids such as β-carotene; and compounds such as plant polyphenols. Fat-soluble vitamin E is a chain-breaking free-radical scavenger and is particularly effective at preventing initiation and propagation of lipid peroxidation in cellular membranes and, thus, maintaining membrane stability. Water-soluble vitamin C is capable of regenerating vitamin E in the antioxidant cascade as illustrated in Figure 8.13.

Polyphenol compounds scavenge peroxyl, superoxide, and nitric oxide radicals. Rich sources of polyphenols include tea, wine (particularly red wine), fruits, and vegetables. Flavonoids, a subclass of polyphenols that includes the flavonols, flavonones, and anthocyanidins contain a number of phenolic hydroxyl ($^-$OH) groups attached to ring structures that confer their antioxidant activity. Differences in the antioxidant power of individual flavonoids result from the variation in the number and arrangement of the hydroxyl groups, as well as extent and nature of the glycolation of these groups. Flavonoids are reported to inhibit the enzyme lipoxygenase and to inhibit oxidant generation by neutrophils in a manner similar to vitamin E. As with vitamin E, the antioxidant power of the flavonoids results from donating a hydrogen atom from a hydroxyl moiety. However, studies in chemical systems show flavonoids to be 3 to 10 times more effective than vitamin E in their scavenging abilities. In addition, they may have anti-inflammatory effects. Flavonoids are used clinically to decrease inflammation and endothelial damage in diseases in which free radicals play a major role. In addition, flavonoids are also scavengers of the nitric oxide radical (NO$^•$). The NO$^•$ radical generated by inflammatory cells is reported to be toxic after reaction with the $O_2^{•-}$ radical. Scavenging NO$^•$ radicals could therefore contribute to the proposed beneficial effect of the flavonoids.

Some immune cell functions can be impaired by an excess of free radicals

The immune system is the body's defence against invading microorganisms that can cause illness and infection, but immune impairment is observed after strenuous exercise lasting more than one hour and some immune cell functions can be impaired by an excess of free radicals. There is some controversy regarding the effects of antioxidant vitamins on immune function and resistance to infections. A few studies have reported fewer infectious episodes in athletes supplemented with vitamin C prior to long-distance foot races, although other studies have not confirmed this finding. There are also conflicting results regarding the effects of antioxidant vitamins on hormonal and immune responses to exercise.

Free radicals may contribute to the development of muscle fatigue

The activity-dependent increase in ROS production appears to contribute to the development of fatigue of skeletal muscle during strenuous exercise. Experimental evidence suggests that muscle-derived ROS primarily act on myofibrillar proteins to inhibit calcium sensitivity and depress force. Decrements in calcium sensitivity and force are acutely reversible by dithiothreitol, a thiol-selective reducing agent. Fatigue causes oxidation of glutathione, a thiol antioxidant in muscle fibres, and is reversed by thiol-specific reducing agents. N-acetylcysteine (NAC), a drug that supports glutathione synthesis, has been shown to lessen oxidation of cellular constituents and delay muscle fatigue both *in vivo* in humans and animals and *in vitro* in isolated muscle preparations. In humans, NAC pretreatment improves performance of limb and respiratory muscles during fatigue protocols and extends time to task failure during volitional exercise. These observations suggest that thiol modifications on one or more regulatory proteins are responsible for oxidant-induced losses of force production during fatigue. More intense ROS exposure leads to losses in calcium regulation that mimic pathologic changes and are not reversible. These findings identify ROS as endogenous mediators of muscle fatigue, and show the feasibility of thiol-based countermeasures. In contrast, RNS depress force production but do not appear to cause fatigue of healthy muscle. Future research may define the cellular mechanism of ROS action and develop antioxidants as novel therapeutic interventions for treating fatigue.

 Leaders in the field

Eric Newsholme began his academic career as a classically trained biochemist and he published a long series of papers (over 100 in total) in the *Biochemical Journal* on the regulation of substrate utilization in muscle. He has made major contributions to our understanding of mechanisms of metabolic regulation in cardiac and skeletal muscle, the metabolic adaptations that take place in response to training, and of the role of nutrients in the regulation of immune cell function. Having taken a strong personal interest in exercise relatively late in life (and having completed many marathons), Eric applied his extensive knowledge of exercise metabolism to the problem of fatigue. He developed a 'central fatigue hypothesis' based on the availability of tryptophan as a metabolic precursor for the neurotransmitter serotonin (5-hydroxytryptamine). He is also the author of several classic textbooks on the regulation of metabolism.

Antioxidant supplements could impair training adaptations

In recent years it has become clear that free radicals may actually have advantageous effects. Both ROS and RNS are involved in the modulation of cell-signalling pathways and the control of several (redox-sensitive) transcription factors. Although high levels of ROS may interfere with muscle function, moderate levels of ROS are actually essential in the development of optimal force production in muscle. These findings raise a number of important questions concerning the possible role of free-radical species as signals for wider adaptive responses in these and other tissues, and question the approach to protection of tissues that involves advocating the use of widespread supplementation with antioxidant nutrients. It is entirely feasible that those adaptations to stress mediated by free radicals play an important role in maintaining cell viability in tissues routinely subjected to repeated stresses (e.g. muscle following exercise) and that increased consumption of some antioxidant nutrients might interfere with these necessary adaptive processes.

In one recent study, fourteen men were trained for eight weeks. Five of the men were supplemented daily with an oral dose of 1 g vitamin C. The administration of vitamin C significantly impaired endurance capacity. The adverse effects of vitamin C may result from its capacity to reduce the exercise-induced expression of key transcription factors involved in mitochondrial biogenesis: peroxisome proliferator-activated receptor co-activator 1, nuclear respiratory factor 1, and mitochondrial transcription factor A. Vitamin C also prevented the exercise-induced expression of cytochrome c (a marker of mitochondrial content) and of the antioxidant enzymes superoxide dismutase and glutathione peroxidase. Thus, it seems that supplementation with large doses of antioxidants may interfere with the function of ROS and may reduce cellular adaptations to exercise training. A lot more research needs to be done to determine the exact role of ROS in training, but it seems that they may not always be damaging and the role of ROS in modulating signalling pathways may have been ignored for a long time.

There appear to be some risks with high-dose antioxidant supplementation

Antioxidants obviously provide some defence against the damaging effects of free radicals and the media commonly extol the potential health benefits of a high antioxidant intake. This has prompted many individuals to take large doses of antioxidant vitamins. However, excessive antioxidant ingestion may not be uniformly helpful. For example, in heavy smokers, increased intake of vitamin E and β-carotene actually increases the incidence of lung cancer. One possible reason for this effect is that antioxidants may interfere with important processes needed to kill cancer cells. Administration of antioxidants inhibits apoptosis (cell death), an important defence mechanism that inhibits tumour development by eliminating new mutated cells. ROS are intermediate messengers in several apoptosis-signalling pathways. So, excess antioxidants may effectively 'shoot the messengers.' Thus, in situations where individuals have damaged their DNA (e.g. by heavy smoking) the administration of antioxidants may prevent the effective removal of damaged cells.

Some side-effects are associated with consuming excessive amounts of individual antioxidant vitamins. Very large doses of vitamin C are associated with urinary stone formation, impaired copper absorption, and diarrhoea. Excess intakes of vitamin A by pregnant women can cause birth defects and recent evidence suggests that high doses of vitamin A may be associated with reduced bone density and increased risk of hip fractures in post-menopausal women. Large intakes of vitamin E can impair absorption of vitamins A and K. Thus, more is not always better.

The controversy continues as to whether physically active individuals should consume large amounts of antioxidant compounds. At present, the data are insufficient to strongly recommend antioxidant supplements for athletes. In the last few decades, the role of ROS in exercise physiology has received considerable attention. Acute exercise has been shown to induce an elevated generation of ROS in skeletal muscle via different mechanisms and there is clear evidence that ROS formation in response to vigorous physical exertion can result in oxidative stress. However, recent research has revealed the important role of ROS as signalling molecules that modulate contractile function and adaptive processes in skeletal muscle. In particular ROS seem to be involved in the modulation of gene expression via redox-sensitive transcription pathways. This potentially represents an important regulatory mechanism, which has been suggested to be involved in the process of training adaptation. In this context, the adaptation of endogenous antioxidant systems in response to regular training reflects a potential mechanism responsible for augmented tolerance of skeletal muscle to exercise-induced stress. If so, it is likely that recommendations to athletes on antioxidant supplements may soon change. Consumption of high-dose supplements could actually impair the ability of the athlete to adapt to the training stimulus and some evidence for this has emerged from recent animal and human studies. Further research is needed to fully document the effects on training adaptation of long-term antioxidant use. The old view that ROS are a byproduct of metabolism and only have a deleterious effect on muscle function is now replaced by the concept that specific ROS are generated in a controlled manner by skeletal muscle fibres in response to physiological stimuli, and play important roles in the physiological adaptations of muscle to contractions. These include optimization of contractile performance and initiation of key adaptive changes in gene expression to the stresses of contractions.

Physical performance is unlikely to benefit from supplementation with complex blends of antioxidants in individuals who are not deficient

Results of studies undertaken to investigate possible benefits of antioxidant supplement consumption have been equivocal, ranging from apparent cardiovascular benefits by consuming vitamin E supplements to apparent damage associated with β-carotene supplementation in smokers. While these studies were well controlled, no intervention study to date has identified any health or physical performance benefits resulting from supplementation with complex blends of antioxidants such as those found in fruits, vegetables, and nuts in individuals who were not experiencing micronutrient deficiency. Moreover, it is clear that simplistic interventions consisting of one or two antioxidant components may be ineffectual, and indeed might be deleterious by disturbing the general balance of the various factors within the oxidant defence system. Given the complexity of human biological mechanisms in place to cope with oxidative challenges, it is unlikely that reliance on supplementation with single antioxidants will be the most effective defence. Therefore, a wide variety of foods, especially fruits, vegetables and nuts, containing complex blends of components important for the oxidant defence system should be consumed. Consumption of appropriately fortified foods may help to ensure that the diverse needs of our body's oxidant defence system are met; but the available evidence does not at this time support the use of supplements. As our knowledge increases, it may be possible to define specific levels and combinations of antioxidants to counter the effects of free radicals and oxidative stress.

Non-steroidal anti-inflammatory drugs and training adaptation

Non-steroidal anti-inflammatory drugs (NSAIDs), including ibuprofen, aspirin (acetylsalicylic acid), naproxen, and diclofenac, are perhaps the most widely known therapy in the treatment of muscle damage. Athletes often take these drugs to relieve the pain or soreness following strenuous exercise. The evidence of their effectiveness is mixed, with some studies showing a reduction in muscle pain and some studies showing no change. Fairly consistently, however, studies have shown a reduction in the creatine kinase (CK) response after damaging exercise. The appearance of elevated levels of CK in plasma is often used as a marker of muscle damage although it does not always correlate very well with delayed onset muscle soreness and other markers of muscle damage such as loss of strength. NSAIDs inhibit the synthesis of certain prostaglandins, which are potential mediators of oedema and pain during acute inflammation. NSAIDs may therefore interfere with the normal inflammatory response after damaging exercise, and it is possible that this inflammatory response plays a role in the adaptation that occurs following exercise. There is some evidence of suppressed tissue protein synthesis following high-intensity eccentric exercise as a result of high doses of ibuprofen and paracetamol. There is also evidence from animal studies that NSAIDs can interfere with muscle regeneration and hypertrophy. Given the equivocal acute effects of NSAIDs and their likely negative effect on training adaptation, NSAIDs should not be recommended as a strategy to treat symptoms of muscle damage. These findings provide more evidence that strategies that interfere with the normal signalling pathways have the potential to reduce the training adaptations and make the training less effective.

KEY POINTS

1. Several principles govern the nature and extent of physiological and metabolic adaptations to training. These include the overload principle, the specificity of exercise principle, and the reversibility principle.

2. Heavy resistance training for several months causes hypertrophy of the muscle fibres, thus increasing total muscle mass and the maximum power output that can be developed.

3. Training for strength, power, or speed has little, if any, effect on aerobic capacity. Both heavy resistance training and sprinting bring about specific changes in the immediate (ATP and PCr) and short-term (glycolytic) energy-delivery systems, and improvements in strength and/or sprint performance.

4. Adaptations to aerobic endurance training include increases in capillary density and mitochondrial size and number in trained muscle. The activity of the TCA cycle and other oxidative enzymes are increased with a concomitant increase in the capacity to oxidize both lipid and carbohydrate.

5. Intramuscular stores of myoglobin, glycogen, and triacylglycerol are increased by endurance training.

6. Endurance training also brings about cardiovascular adaptations including enhanced blood volume, stroke volume, and cardiac output, and an expanded arteriovenous oxygen difference.

7. During exercise endurance-trained muscle exhibits lower rates of carbohydrate utilization and lactate production, whereas lipid oxidation rates are increased compared with untrained muscle. The main source of the increased lipid oxidation after training appears to be from intramuscular triacylglycerol.

8. Endurance training attenuates the magnitude of the hormonal responses to acute exercise.

9. Training adaptations in skeletal muscle may be generated by the cumulative effects of transient increases in gene transcription during recovery from repeated bouts of exercise. Stretch, contraction, and damage of muscle fibres during exercise provide the stimuli for adaptation, which involves changes in the expression of genes coding for different myosin isoforms and enzymes of energy metabolism.

10. Signal-transduction pathways control the transcription of genes into mRNA, translation of mRNA into protein, protein modification altering catalytic activity, regulation of protein degradation, and regulation of cell division, proliferation and fusion.

11. AMPK is activated by a high AMP/ATP ratio and plays a crucial role in energy metabolism acting as a metabolic master switch regulating several intracellular systems including the cellular uptake of glucose, the β-oxidation of fatty acids, and the synthesis of glucose transporter 4 (GLUT4) and biogenesis of mitochondria. Insulin and insulin-like growth factor pathways play an important role in muscle hypertrophy through the Akt-mTOR signalling cascade.

12. One family of transcriptional regulators in particular, the peroxisome proliferator-activated receptor γ coactivator family (PGCs) are important in driving mitochondrial biogenesis.

13. If the training load is too high with insufficient time for recovery between sessions, the beneficial adaptations that are sought will not occur, and instead there will be a loss of functional capacity. This is often referred to as overtraining, a condition in which underperformance is experienced despite continued or even increased training.

14. Acute prolonged bouts of strenuous exercise and periods of heavy training appear to increase the susceptibility of the athlete to infections due to a diminution of immune system function.

15. Elevated levels of stress hormones, particularly glucocorticoids such as cortisol, are probably mostly responsible for causing temporary immunosuppression. This may be mediated via a glucocorticosteroid-induced inhibition of IFN-γ production by T lymphocytes. When exercise is repeated frequently there may not be sufficient time for the immune system to recover fully.

16. The diet during training may modify metabolic adaptations.

17. Low levels of muscle glycogen and high-fat diets may enhance the mRNA content of some genes involved in exercise metabolism although whether this can also result in improvements in performance is not clear.

18. Increased free-radical generation occurs during exercise, but the evidence for beneficial effects of nutritional antioxidant supplementation is equivocal. Antioxidant supplementation may decrease the training efficiency by preventing some cellular adaptations to exercise.

Selected further reading

Akerstrom TCA *et al.* (2009). Glucose ingestion during endurance training does not alter adaptation. *American Journal of Applied Physiology* 106: 1771–1779.

Atherton PJ *et al.* (2005). Selective activation of AMPK-PGC-1alpha or PKB-TSC2-mTOR signalling can explain specific adaptive responses to endurance or resistance training-like electrical muscle stimulation. *FASEB Journal* 19: 786–788.

Baar K *et al.* (2002). Adaptations of skeletal muscle to exercise: rapid increase in the transcriptional coactivator PGC-1. *FASEB Journal* 16: 1879–1886.

Bickham DC *et al.* (2006). The effects of short-term sprint training on MCT expression in moderately endurance-trained runners. *European Journal of Applied Physiology* 96: 636–643.

Cameron-Smith D *et al.* (2003). A short-term, high-fat diet up-regulates lipid metabolism and gene expression in human skeletal muscle. *American Journal of Clinical Nutrition* 77: 313–318.

Coffey VG *et al.* (2006). Early signalling responses to divergent exercise stimuli in skeletal muscle from well-trained humans. *FASEB Journal* 20: 190–192.

De Bock K *et al.* (2008). Effect of training in the fasted state on metabolic responses during exercise with carbohydrate intake. *Journal of Applied Physiology* 104: 1045–1055.

Drummond MJ *et al.* (2009). Nutritional and contractile regulation of human skeletal muscle protein synthesis and mTORC1 signalling. *Journal of Applied Physiology* 106: 1374–1384.

Fujii N *et al.* (2000). Exercise induces isoform-specific increase in 5′AMP-activated protein kinase activity in human skeletal muscle. *Biochemical and Biophysical Research Communications* 273: 1150–1155.

Gleeson M (2005). *Immune function in sport and exercise.* Edinburgh: Elsevier.

Gleeson M (2007). Exercise and immune function. *Journal of Applied Physiology* 103: 693–699.

Gleeson M and Bishop NC (1999). Immunology. In: *Basic and applied sciences for sports medicine* (ed. Maughan RJ). Oxford: Butterworth Heinemann, pp. 199–236.

Gomez-Cabrera MC *et al.* (2005). Decreasing xanthine oxidase-mediated oxidative stress prevents useful cellular adaptations to exercise in rats. *Journal of Physiology* 567: 113–120.

Gomez-Cabrera MC *et al.* (2008). Oral administration of vitamin C decreases muscle mitochondrial biogenesis and hampers training-induced adaptations in endurance performance. *American Journal of Clinical Nutrition* 87: 142–149.

Halson SL *et al.* (2002). Time course of performance changes and markers of fatigue during intensified training in cyclists. *Journal of Applied Physiology* 93: 947–956.

Hansen AK *et al.* (2005). Skeletal muscle adaptation: training twice every second day vs. training once daily. *Journal of Applied Physiology* 98: 93–99.

Hargreaves M and Cameron-Smith D (2002). Exercise, diet, and skeletal muscle gene expression. *Medicine and Science in Sports and Exercise* 34: 1505–1508.

Hawley JA (2002). Training for enhancement of sports performance. In: *Physiological bases of sports performance* (ed. Hargreaves M and Hawley J). Sydney: McGraw-Hill Australia, pp. 125–151.

Hawley JA, Tipton KD, and Millard-Stafford ML (2006). Promoting training adaptations through nutritional interventions. *Journal of Sports Sciences* 24: 709–721

Holloszy JO (2008). Regulation by exercise of skeletal muscle content of mitochondria and GLUT4. *Journal of Physiology and Pharmacology* 59(Suppl 7): 5–18.

Jackson MJ (2007). Free radicals generated by contracting muscle: By-products of metabolism or key regulators of muscle function? *Free Radical Biology and Medicine* 44: 132–141.

Ji LL (2007). Antioxidant signalling in skeletal muscle: a brief review. *Experimental Gerontology* 42: 582–593.

Jones Jump DB and Clarke SD (1999). Regulation of gene expression by dietary fat. *Annual Review of Nutrition* 19: 63–90.

Kuipers H and van Breda E (2002). Overtraining. In: *Physiological bases of sports performance* (ed. Hargreaves M and Hawley J). Sydney: McGraw-Hill Australia, pp. 108–124.

Lee-Young RS *et al.* (2006). Carbohydrate ingestion does not alter skeletal muscle AMPK signalling during exercise in humans. *American Journal of Physiology (Endocrinology and Metabolism)* 291: E566–573.

Matsakas A and Patel K (2009). Intracellular signalling pathways regulating the adaptation of skeletal muscle to exercise and nutritional changes. *Histology and Histopathology* 24: 209–222.

McGee SL *et al.* (2003). Exercise increases nuclear AMPK alpha2 in human skeletal muscle. *Diabetes* 52: 926–928.

Narkar VA *et al.* (2008). AMPK and PPARdelta agonists are exercise mimetics. *Cell* 134: 405–415.

Niess AM and Simon P (2007). Response and adaptation of skeletal muscle to exercise – the role of reactive oxygen species. *Frontiers in Bioscience* 12: 4826–4838.

Pedersen BK and Bruunsgaard H (1995). How physical exercise influences the establishment of infections. *Sports Medicine* 19: 393–400.

Peters SJ *et al.* (2001). Human skeletal muscle PDH kinase activity and isoform expression during three days of a high fat/low carbohydrate diet. *American Journal of Physiology (Endocrinology and Metabolism)* 281: E1151–E158.

Petitbois C *et al.* (2002). Biochemical aspects of overtraining in endurance sports. *Sports Medicine* 32: 867–878.

Pilegaard H *et al.* (2002). Influence of pre-exercise muscle glycogen content on exercise-induced transcriptional regulation of metabolic genes. *Journal of Physiology* 541.1: 261–271.

Pilegaard H *et al.* (2003). Exercise induces transient transcriptional activation of the PGC-1alpha gene in human skeletal muscle. *Journal of Physiology* 546: 851–858.

Powers SK *et al.* (2004). Dietary antioxidants and exercise. *Journal of Sports Sciences* 22: 81–94.

Powers SK and Jackson MJ (2008). Exercise-induced oxidative stress: cellular mechanisms and impact on muscle force production. *Physiological Reviews* 88: 1243–1276.

Reid MB (2008). Free radicals and muscle fatigue: Of ROS, canaries, and the IOC. *Free Radical Biology and Medicine* 44: 169–179.

Sen CK *et al.* (2000). Exercise-induced oxidative stress and antioxidant nutrients. In *Nutrition in sport* (ed. Maughan RJ). Oxford: Blackwell Science, pp. 292–317.

Steensberg A *et al.* (2000). Production of interleukin-6 in contracting human skeletal muscles can account for the exercise-induced increase in plasma interleukin-6. *Journal of Physiology* 529: 237–242.

Steensberg *et al.* (2007). Nitric oxide production is a proximal signalling event controlling exercise-induced mRNA expression in human skeletal muscle. *FASEB Journal* 21: 2683–2694.

St-Pierre J *et al.* (2002). Topology of superoxide production from different sites in the mitochondrial electron-transport chain, *Journal of Biological Chemistry* 277: 44784–44790.

Trappe TA *et al.* (2002). Effect of ibuprofen and acetaminophen on postexercise muscle protein synthesis. *American Journal of Physiology (Endocrinology and Metabolism)* 282: E551–556.

Tunstall RJ *et al.* (2002). Exercise training increases lipid metabolism gene expression in human skeletal muscle. *American Journal of Physiology (Endocrinology and Metabolism)* 283: E66–E72.

Wackerhage H and Woods NM (2002). Exercise-induced signal transduction and gene regulation in skeletal muscle. *Journal of Sports Science and Medicine* 1: 103–114.

Wilkinson SB *et al.* (2008). Differential effects of resistance and endurance exercise in the fed state on signalling molecule phosphorylation and protein synthesis in human muscle. *Journal of Physiology* 586: 3701–3717.

Williams RS and Neufer PD (1996). Regulation of gene expression in skeletal muscle by contractile activity.

In: *Handbook of physiology*, Section 12: *Exercise: regulation and integration of multiple systems* (ed. Rowell LB and Shepherd JT). New York: Oxford University Press, pp. 1124–1150.

Winder WW and Hardie DG (1996). Inactivation of acetyl-CoA carboxylase and activation of AMP-activated protein kinase in muscle during exercise. *American Journal of Physiology* 270: E299–304.

Xia R *et al.* (2003). Skeletal muscle sarcoplasmic reticulum contains a NADH-dependent oxidase that generates superoxide, *American Journal of Physiology* 285: C215–C221.

Yang Y *et al.* (2005). Time course of myogenic and metabolic gene expression in response to acute exercise in human skeletal muscle. *Journal of Applied Physiology* 98: 1745–1752.

Yeo WK *et al.* (2008). Skeletal muscle adaptation and performance responses to once a day versus twice every second day endurance training regimens. *Journal of Applied Physiology* 105: 1462–1470.

Appendix 1
Key concepts in physical, organic, and biological chemistry

Introduction

The study of exercise biochemistry requires an understanding of some simple physical and organic chemistry. Carbohydrates, lipids, proteins, and nucleic acids are themselves composed of smaller building blocks. This appendix contains a short review of important chemical concepts, interactions, and processes involving biomolecules, together with a brief summary of membrane transport mechanisms and the structure and function of the various cellular organelles.

Matter, energy, and atomic structure

The human body consists only of matter and energy in their various forms. Indeed, the same can be said for the entire universe. Matter occupies space and has a mass that represents the quantity of matter that is present. We often equate mass with weight, which is the force that gravity exerts upon the mass, but technically there is a difference. The quantity or mass of an object does not change regardless of its location, but its weight varies according to the pull of gravity. For example, a rock weighing 6 kg on earth would weigh only about 1 kg on the surface of the moon because the gravitational force of the moon is only about one-sixth of that of the earth.

Energy is the capacity of any system, including the living body, to do work, that is to produce a change of some sort in matter. Energy can exist in several forms including light, heat, electrical, mechanical, and chemical energy. Different forms of energy can be transformed from one form to another. In the body the potential chemical energy stored in foodstuffs can be transformed to do various forms of work such as movement or the synthesis of large molecules from small molecules. Matter and energy are interrelated by Einstein's famous equation:

$$E = mc^2,$$

where E is energy, m is mass, and c is the speed of light (about 299 744 km/s). Nothing is capable of moving faster than light. Einstein's equation can, in principle, go in either direction. Thus, energy can be transformed into matter, and matter can be transformed into energy.

There are many thousands of different types of matter, but all can be reduced into smaller units. The smallest units into which a substance can be broken down chemically are the elements, each of which has different and unique properties. Ninety-two different elements are presently known to exist, but only about twelve are common in living organisms. The most abundant are oxygen, carbon, hydrogen, and nitrogen (in that order). These four elements comprise 96% of the mass of a human.

Atoms are the smallest units of an element that retain all the properties of the element. The atoms of all elements can be broken down physically into the same subatomic particles: protons, neutrons, and electrons. Hence, the atoms of the various elements only differ in the numbers of protons, neutrons, and electrons that they contain. **Protons**, which possess a positive charge, and **neutrons**, which are electrically neutral, are held together to form the nucleus of an atom. **Electrons**, which have negligible mass (only about one-eight thousandth of that of a proton or a neutron) and possess a negative charge, spin around the atomic nucleus in discrete **orbitals**, which may be spherical or dumb-bell shaped. Some electrons move in orbitals close to the nucleus, others farther away. Those moving in orbits farther from the nucleus have more energy than those close to the nucleus, so, in a sense, the orbitals can be thought of as energy levels. Electrons normally stay at their particular energy level, but by gaining or losing energy they can jump from one energy level or orbital to another. Although electron orbitals can vary in shape, electrons at each energy level can be depicted diagrammatically as moving in concentric and circular orbits or shells around the nucleus. A maximum of two electrons can be held in the innermost shell. The second and third shells can hold up to eight electrons each. The fourth shell can hold up to 18 electrons.

The number of electrons equals the number of protons in the nucleus, resulting in an atom that is electrically neutral. The smallest atom is that of hydrogen, which is composed of just one proton and one electron (Figure A1.1a). The carbon atom consists of six protons, six neutrons, and six electrons, whereas an oxygen atom is made up of eight protons, eight neutrons, and eight electrons. Both oxygen and carbon have two electrons in their inner shells, but differ in the number in the second shell; oxygen has six electrons in this shell, and carbon four (Figure A1.1b). The chemical properties of an element and the way it reacts with other elements depend on the number of electrons in its outer shell. If this shell is full, the element does not react with others and is said to be inert. Helium, with two electrons, and neon, with 18 electrons, are examples of inert elements. Atoms whose outer shell is not full tend to move towards a more stable configuration by losing, gaining, or sharing electrons with other atoms. The atoms become bound together by attractive forces called **chemical bonds**. These represent a source of potential chemical energy. Breaking chemical bonds releases some energy that can be used to do useful work.

Nearly all the mass of an atom is in its protons and neutrons, so their combined number is a measure of the **atomic mass** of the element. An element's mass is indicated by a superscript in front of the element, e.g. ^1H, ^{12}C, ^{16}O. The number of protons in an atom is its **atomic number** and this is indicated as a subscript in front of the symbol for the atom, e.g. ^1_1H, $^{12}_6C$, $^{16}_8O$.

All of the atoms of a given element contain the same number of protons and electrons (which determines the chemical properties of the element), but in some elements the number of neutrons in the nucleus, and hence the atomic mass (but not the atomic number), varies. Atoms that have the same atomic number but differ in the number of neutrons are called **isotopes**. Some isotopes are unstable and emit radiation in the form of gamma rays, electrons, or a helium nucleus (two protons and two neutrons). This radiation can be measured and some unstable isotopes have proved useful as tracers. For example, normal carbon has an atomic mass of 12, but the **radioactive isotope** of carbon has a mass of 14. Glucose, a simple sugar, can be prepared using $^{14}_6C$ instead of $^{12}_6C$ and its metabolism in the body can be followed by identifying the presence of $^{14}_6C$ in intermediate compounds and expired carbon dioxide. Nowadays, it is also possible to detect the presence and quantity of **stable isotopes** (ones that do not decay by the emission of radiation), such as $^{13}_6C$, and these are increasingly being used in place of the more dangerous radioactive isotopes in metabolic studies in humans.

(a) Hydrogen atom
      (1p)

electron

proton

● proton
○ neutron
● electron

(b) Carbon atom
      (6p, 6n)

(c) Oxygen atom
      (8p, 8n)

Figure A1.1

Diagrammatic representation of atoms of (a) hydrogen, (b) carbon, and (c) oxygen.

Figure A1.2

Diagrammatic representation of a molecule of water showing the distribution of partial charges.

In the body, most elements are not present as free atoms, but are combined with others to form **molecules**. For example, a molecule of water contains one atom of oxygen bound to two atoms of hydrogen, symbolized as H_2O (Figure A1.2). Even the oxygen in the air we breathe is not made up of free atoms of oxygen, but of molecular oxygen consisting of two atoms of oxygen bound together and symbolized as O_2. A molecule of a simple hexose sugar, glucose, contains 24 atoms, six of carbon, 12 of hydrogen and six of oxygen, and this formula can be expressed as $C_6H_{12}O_6$; this is known as the empirical chemical formula. The **molecular mass** is obtained by adding up the atomic masses present in the molecule. Thus, for glucose we have $6 \times {}^{12}C + 12 \times {}^{1}H + 6 \times {}^{16}O = 72 + 12 + 96 = 180$. The molecular mass of a substance in grams is equal to one mole of that substance and the number of molecules in a **mole** of any substance is the same and is known as **Avogadro's number** and equals 6.022×10^{23}.

Many substances in the body are dissolved in water and it is useful to have some measure of the concentration of molecules in solution. One mole of a substance dissolved in enough water to make one litre is known as a **one molar solution** (1.0 M). Biochemists usually deal with smaller concentrations such as millimoles per litre (1 mM = 1×10^{-3} M).

Chemical bonds, free energy, and ATP

This section provides a brief review of chemical bonds. Emphasis is placed on bonds between the six major elements found in the living body: hydrogen (H), carbon (C), nitrogen (N), oxygen (O), phosphorus (P), and sulphur (S).

Covalent bonds hold together two or more atoms by the interaction of their outer electrons. Covalent bonds are the strongest chemical bonds. The energy of a typical covalent bond is approximately 330 kilojoules per mole (kJ/mol). However, this can vary from about 210 to around 460 kJ/mol depending on the elements involved. Once formed, covalent bonds rarely break spontaneously. This is due to simple energetic considerations; the thermal energy of a molecule

Bond number	Example	Energy (kJ/mol)

single

$$H - \underset{\underset{H}{|}}{\overset{\overset{H}{|}}{C}} - H \qquad 330$$

double

$$H - \underset{\underset{H}{|}}{\overset{\overset{H}{|}}{C}} = \underset{\underset{H}{|}}{\overset{\overset{H}{|}}{C}} - H \qquad 630$$

triple

$$H - C \equiv C - H \qquad 840$$

Figure A1.3

Single, double, and triple covalent bonds.

at room temperature (20°C or 293 K) is only about 2.5 kJ/mol, much lower than the energy required to break a covalent bond. There are single, double, and triple covalent bonds (Figure A1.3).

Carbon–carbon bonds are particularly strong and stable covalent bonds. The major organic elements have standard bonding capabilities: C, N, and P can form up to four covalent bonds with other atoms, O and S can form two and H can only form one. In solution, O and S can lose a proton (or hydrogen ion, H^+), leaving the O or S with a negative charge (Figure A1.4). Covalent bonds can have partial charges when the atoms involved have different electronegativities. Water is an obvious example of a molecule with partial charges. The symbols delta-plus ($\delta+$) and delta-minus ($\delta-$) are used to indicate partial charges (Figure A1.2). Oxygen, because of its high electronegativity, attracts the electrons away from the hydrogen atoms, resulting in a partial negative charge on the oxygen and a partial positive charge on each of the hydrogens. The possibility of hydrogen bonds is a consequence of these partial charges.

Hydrogen bonds (H-bonds) are weak intra- or intermolecular attractions between a hydrogen atom and an electronegative atom possessing a lone pair of electrons (e.g. oxygen or nitrogen atoms). Hydrogen bonds are formed when a hydrogen atom is shared between two molecules (Figure A1.5 shows the hydrogen bonding between water molecules). Hydrogen bonds have polarity. A hydrogen atom covalently attached to a very electronegative atom (N, O, or P) shares its partial positive charge with a second electronegative atom (N, O, or P). Hydrogen bonds are 21 kJ/mol in strength. These bonds are frequently found in proteins and

$$H - \underset{\underset{H}{|}}{\overset{\overset{H}{|}}{C}} - \underset{\underset{OH}{|}}{\overset{\overset{H}{|}}{C}} - \overset{\overset{O}{||}}{C} - OH \quad \rightleftharpoons \quad H - \underset{\underset{H}{|}}{\overset{\overset{H}{|}}{C}} - \underset{\underset{OH}{|}}{\overset{\overset{H}{|}}{C}} - \overset{\overset{O}{||}}{C} - O^- \quad + \quad H^+$$

Lactic acid   Lactate anion   Hydrogen ion (proton)

Figure A1.4

Dissociation of a proton (H^+ ion) from lactic acid leaving an oxygen (O) with a negative charge.

Figure A1.5

Hydrogen bonding between water molecules. Hydrogen bonds have polarity. A hydrogen atom covalently attached to a very electronegative atom (N, O, or P) shares its partial positive charge with a second electronegative atom (N, O, or P).

nucleic acids, and by reinforcing each other serve to keep the protein (or nucleic acid) structure secure. However, as the hydrogen atoms in the protein could also hydrogen bond to the surrounding water, the relative strength of protein–protein H-bonds versus protein–H_2O bonds is smaller than 21 kJ/mol.

Ionic bonds are formed when there is a complete transfer of electrons from one atom to another. The valence (outer shell) electrons are either lost or gained, resulting in two ions, one positively charged and the other negatively charged. The ions that are oppositely charged are held together by electrostatic forces. For example, when a sodium atom (Na) donates the one electron in its outer valence shell to a chlorine atom (Cl), which needs one electron to fill its outer valence shell, NaCl (salt) results. The symbol for sodium chloride is Na^+Cl^-. Ionic bonds are 17–30 kJ/mol in strength.

van der Waals interactions are very weak bonds, generally no greater than 5 kJ/mol, formed between non-polar (uncharged) molecules or non-polar parts of a molecule. The weak bond is created because a C–H bond can have transient dipole and can induce a transient dipole in another C–H bond.

Non-polar molecules cannot form H-bonds with water, and are therefore very poorly soluble in water. These molecules are known as **hydrophobic** (water-hating), as opposed to **hydrophilic** (water-loving) molecules, which can form H-bonds with water. Hydrophobic molecules tend to aggregate together in avoidance of water molecules; **hydrophobic interactions** are clearly demonstrated when a drop of oil is placed on water. The oil forms a thin layer over the surface of the water. If mixed together vigorously, the oil forms small globules but does not dissolve in the water. This attraction/repulsion is known as the hydrophobic force. To understand the energetics driving this interaction, visualize the water molecules surrounding a single 'dissolved' molecule attempting to form the greatest number of H-bonds with each other. The best energetic solution involves forcing all of the non-polar molecules together, thus reducing the total surface area that breaks up the water's H-bond matrix.

Breaking chemical bonds releases energy: during biochemical reactions that involve the breaking of chemical bonds, energy is released, some of which is unavailable to do work (entropy). In addition, **free energy** and heat energy (enthalpy) also become available. The free energy released, usually signified by the symbol G, can be used to do useful work in the body. Free energy released during the breakdown (catabolism) of carbohydrates and fats can be stored in the compound adenosine triphosphate (ATP).

ATP is the only form of chemical energy that can be converted into other forms of energy used by living cells. The energy released from the breakdown of ATP to ADP and inorganic phosphate (P_i) can be used in the biosynthesis of macromolecules (anabolism) or for other energy-requiring processes such as active transport or mechanical work.

Reactions involving the breaking of phosphate bonds and the liberation of P_i are catalysed by enzymes called **kinases**. In the case of ATP breakdown, these are commonly abbreviated to ATPases.

All biochemical reactions are **inefficient**, which means that not all the energy released can be conserved or used to do work: some energy is always lost as **heat**. This is used to maintain body temperature at about 37 °C. During exercise, when the rate of catabolic reactions is markedly increased in the active muscles to provide energy for contraction, the rate of heat production also increases substantially and muscle temperature rises by 1–5 °C.

Because of its unique role in energy metabolism, ATP has been termed the energy currency of the cell. However, unlike money, ATP cannot be accumulated in large amounts and the **intramuscular ATP** concentration is only about 5 mmol/kgww. During maximal exercise, there is only sufficient ATP present to fuel about 2 s of muscle force generation. Experiments have shown that the muscle ATP store never becomes completely depleted because it is normally efficiently resynthesized from ADP and P_i at the same rate at which it is degraded.

During submaximal steady-state exercise ATP resynthesis is achieved by mitochondrial oxidation of carbohydrates and lipids. This is commonly referred to as **aerobic** metabolism, because it involves the use of oxygen. However, at the onset of exercise and in high-intensity exercise this is achieved principally by **anaerobic** (without the use of oxygen) ATP resynthesis.

Hydrogen ion concentration and buffers

An **acid** is a compound able to donate a hydrogen ion (H^+), e.g. HCl, H_2CO_3. A **base** is a compound able to accept a hydrogen ion, e.g. NaOH, HCO_3^-.

The pH is a measure of the concentration of hydrogen ions. These are derived, for example, from the dissociation of an acid, such as HCl, when it is dissolved in water (HCl → H^+ + Cl^-). The pH value is defined as the negative decimal logarithm of the free H^+ concentration or $[H^+]$, i.e. **pH = $-\log_{10}[H^+]$**, where $[H^+]$ is expressed in moles per litre (mol/l or M), so the concentration of free H^+ increases 10-fold for each decrease of 1 pH unit.

The $[H^+]$ in pure water is 10^{-7} mol/l. Therefore, the pH of pure water is:

$$pH = -\log_{10}(10^{-7}) = -(-7) = 7.$$

pH 7 is often referred to as neutral pH. Everything below pH 7 has a higher concentration of H^+ and is considered acidic. Everything above pH 7 has a lower concentration of H^+ and is considered basic; you can also think of this as a higher concentration of OH^-. Most body fluids are close to neutral pH. For example, blood plasma has a pH of 7.4 and resting muscle intracellular fluid (sarcoplasm) has a pH of 7.0.

Buffers act as a reservoir for hydrogen ions and so limit, or buffer, changes in free H^+ concentration. Buffering is passive and almost instantaneous. A buffer consists of a weak acid and its conjugate base. In solution, some but not all molecules of the weak acid dissociate into the conjugate base and hydrogen ions:

$$BH \rightarrow B^- + H^+,$$

where BH is the weak acid and B^- is its conjugate base. The relationship between the concentrations of weak acid, base, and hydrogen ions can be expressed as an equation:

$$K_a[BH] = [H^+][B^-],$$

where K_a is the dissociation constant of the acid, that is the concentration of H^+ at which the concentrations of the base and acid are equal. The equation can be rearranged to give the Henderson equation:

$$[H^+] = K_a[BH]/[B^-],$$

or, alternatively, by taking the negative logarithm of each side of the equation, it becomes the Henderson–Hasselbalch equation:

$$-\log_{10}[H^+] = -\log_{10}K_a + \log_{10}[B^-]/[BH],$$

which is the same as:

$$\textbf{pH} = \textbf{p}K_a + \textbf{log}_{10}[\textbf{B}^-]/[\textbf{BH}],$$

which tells us that **optimal buffering** (minimum change in pH when H^+ ions are added) occurs at pH = pK_a and that this occurs when the concentrations of weak acid (BH) and base (B^-) are equal.

Important buffers in the body are: carbonic acid/bicarbonate (H_2CO_3/HCO_3^-), which is the most important extracellular buffer (with a pK_a of 6.1); haemoglobin and other proteins and peptides containing the amino acid histidine (with pK_a usually 6.0–8.0); and phosphates ($HPO_4^{2-}/H_2PO_4^-$), whose pK_a is 6.8. Note that because the pH of most body fluids is about 7.4, buffers with a pK_a of between 6 and 8 are the most effective.

In exercising muscle large amounts of hydrogen ions are produced from the dissociation of **lactic acid** to H^+ and the lactate anion (Figure A1.4). The resting intracellular pH of muscle is 7.0; a large fall in pH is undesirable as it would denature enzymes and is prevented by the presence of **intracellular buffers** including carnosine (a dipeptide consisting of β-alanine and histidine), phosphocreatine, phosphates, and histidine-containing proteins. As the hydrogen ions diffuse out of the muscle into the blood they are buffered by bicarbonate (in plasma) and haemoglobin (in the erythrocytes). Note that an increase in the blood lactic acid concentration of 10 mmol/l only changes the blood pH by about 0.1; this concentration of lactic acid dissolved in water alone would cause the pH to drop from 7.0 to 2.0. Clearly, the body's buffer systems are very effective!

Membrane structure and transport

Cell membranes are composed of a lipid **bilayer** containing mostly phospholipids and some cholesterol. Within this sea of lipid float proteins, some of which are restricted to one side of the bilayer and some of which are embedded right across from one side of the membrane to the other (Figure A1.6). These proteins have important structural roles, some acting as receptors, channels, or transporters.

Dissolved substances (solutes) move across these semipermeable membranes by simple diffusion, facilitated diffusion, or active transport. Osmosis refers to the movement of water across membranes.

Solutes only move from high to low concentration by **diffusion**. **Simple diffusion** involves the movement of the solute across the lipid bilayer, and hence is to a large extent influenced by the solubility of the substance in lipid. Most water-soluble substances (e.g. glucose) and charged particles (e.g. sodium ions) are poorly soluble in lipid. Large molecules, such as proteins, will not pass across membranes. Very small molecules (e.g. O_2, CO_2, H_2O, NH_3) diffuse easily across cell membranes. Rates of diffusion are affected by temperature and the concentration difference on either side of the membrane.

In **facilitated diffusion** solutes move only from high to low concentration by diffusion but use a specific protein carrier molecule to pass through the membrane. The protein may be a mobile carrier or a gated channel (Figure A1.7). The rate of transport across the membrane is far greater than by simple diffusion and becomes saturated at high concentrations of solute (Figure A1.8). The maximum rate of transport is limited by the number of protein transporters present in the membrane. In muscle, glucose is transported into the fibres by a protein carrier called GLUT4. The number of GLUT4 transporters in the muscle fibre membrane is influenced by exercise and by the hormone insulin.

In **active transport**, substances can be moved against their concentration gradient using a specific protein carrier and **energy** supplied directly by ATP or indirectly by ion electrochemical gradients (e.g. the sodium/potassium pump or Na^+–K^+-ATPase).

The Na^+–K^+-ATPase transports three Na^+ ions out of the cell and two K^+ ions into the cell per molecule of ATP hydrolysed to ADP (Figure A1.9). Because this means that more positively charged ions are moved out of the cell than are transported in, an electrochemical gradient is established. The inside of the cell becomes negative relative to the outside—the difference in electrical potential is about 70 mV in most cells. Inside the cell the Na^+ concentration is maintained at

Figure A1.6

(a) Two-dimensional cross-section of a cell membrane. (b) Three-dimensional view of a cell membrane.

Figure A1.7

Diffusion and facilitated diffusion of substances across cell membranes. In both cases, net movement of the solute molecules occurs from a high concentration of solute to a low concentration of solute.

Figure A1.8

Relative rates of glucose uptake into cells by simple diffusion and facilitated diffusion using a protein transporter in relation to the glucose concentration of the extracellular fluid. Note the much higher rate of uptake with facilitated diffusion compared with simple diffusion, but that at high glucose concentrations, facilitated transport exhibits saturation kinetics. The maximum speed of facilitated transport (V_{max}), indicated by the dashed line, is limited by the number of protein transporters available in the membrane. When all the transporters are full, additional movement can only occur by simple diffusion.

Figure A1.9

The sodium–potassium-ATPase as an example of active transport. Energy released from the hydrolysis of ATP is used to move sodium and potassium ions across the cell membrane against the prevailing concentration gradients. For each molecule of ATP hydrolysed to ADP, three sodium ions are transported out of the cell and two potassium ions are transported in. The presence of separate sodium and potassium ion channels in the membrane is also shown. When these are open, the ions move by diffusion from high to low concentration. The selective opening of sodium channels (allowing a rapid influx of positively charged Na^+ ions) can cause a temporary change in the resting membrane potential. This is called depolarization of the membrane and is important in the generation and propagation of action potentials in excitable cells such as nerve and muscle.

Figure A1.10

Co-transport (symport) of glucose and sodium from the gut lumen into epithelial cells of the small intestine, followed by separate transport across the basal membrane into the blood by the action of a glucose transporter (glucose permease) and the sodium–potassium-ATPase. The sodium–potassium-ATPase generates a large concentration difference for sodium across the membrane. The glucose–sodium symport protein uses that sodium gradient to transport glucose (against its concentration gradient) into the cell.

about 12 mM, while outside it is about 145 mM. The K^+ concentration inside the cell is about 155 mM, whereas outside it is only 4 mM. **Symport** (co-transport) mechanisms use this electrochemical gradient to facilitate the transport of a substance against its concentration gradient. For example, the inward transport of dietary glucose from the gut lumen into intestinal epithelial cells is coupled to that of Na^+, as depicted in Figure A1.10.

Water can readily diffuse across membranes both through the lipid bilayer and through protein pores or channels in the membrane. **Osmosis** is the net movement of water as a consequence of a total solute particle concentration difference across a membrane. Water moves across a semipermeable membrane from a region of low total solute particle concentration (osmolality) to a region of high total solute particle concentration, until the total solute particle concentration is equal on each side of the membrane (Figure A1.11) or its movement is counteracted by the build-up of hydrostatic pressure.

Figure A1.11

Possible mechanisms for permeation of cell membranes by water. Water is a small molecule and may permeate through the spaces between hydrophobic lipid molecules, specific water pores, or other pores (e.g. ion channels). Movement of water molecules is always in the direction of a higher solute (dissolved particle) concentration.

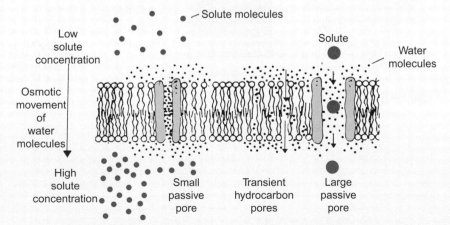

Cells and organelles

All tissues in the body are formed from cells. A typical cell is shown in Figure A1.12. The average diameter of a cell in the human body is about 10 μm (one-hundredth of a millimetre) although there is a wide variety of cell shapes and sizes. Inside the cell is compartmentalized and the organelles are distinct sub-cellular structures. The **nucleus** is the largest organelle; it is usually round or oval-shaped and surrounded by a nuclear envelope composed of two phospholipid membranes. This envelope contains nuclear pores through which messenger molecules pass to the cytoplasm. The nucleus stores and transmits genetic

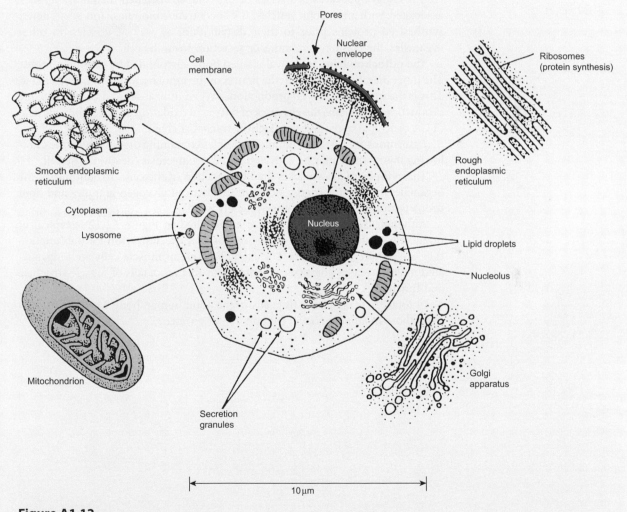

Figure A1.12

Diagram of a typical cell in cross-section, showing the structure of several organelles.

information in the form of deoxyribonucleic acid (DNA). Genetic information passes from the nucleus to the cytoplasm, where amino acids are assembled into proteins. The **nucleolus** is a densely staining region of the nucleus where information concerning ribosomal proteins is being expressed.

The **rough granular endoplasmic reticulum** is an extensive network of folded, sheet-like membranes that has ribosomes attached to its surface. Proteins are synthesized on the ribosomes. The **smooth (agranular) endoplasmic reticulum** is a highly branched tubular network that does not have attached ribosomes. It contains enzymes for fatty acid synthesis and stores and releases calcium; in muscle this is important in the regulation of contraction. The specialized smooth endoplasmic reticulum in muscle is called the sarcoplasmic reticulum.

The **Golgi apparatus** is a series of cup-shaped flattened membranous sacs, associated with numerous vesicles. It concentrates, modifies, and sorts newly synthesized proteins prior to their distribution, by way of vesicles, to other organelles, the plasma membrane, or secretion from the cell.

The **mitochondrion** is an oval-shaped body surrounded by two membranes. The inner membrane folds into the matrix of the mitochondrion, forming cristae. This is the major site of ATP production, oxygen utilization, and carbon dioxide production. It contains the enzymes of fatty acid oxidation, the tricarboxylic acid (TCA or Krebs) cycle, and the electron-transport chain.

Lysosomes are small membranous vesicles containing digestive enzymes. Following injury, these may be activated and cause necrosis (death) of the cell.

The **cytoplasm** or cytosol is the fluid portion of the cell surrounding all the organelles. It contains energy stores in the form of glycogen granules and lipid droplets and the enzymes of anaerobic glycolysis.

Although the cell shape and size as depicted in Figure A1.12 is typical of many cells of the body, other cells show distinct **specialization** for the function they are required to perform. For example, **skeletal muscle cells** are long, striated multinucleated fibres (see Chapter 2 for further details of muscle structure and function).

In total, there are about 10^{14} cells in the adult human body. Most cells (apart from those in adipose tissue) contain 70–80% water.

Appendix 2
Glossary of abbreviations and biochemical terminology

Abbreviations

A	Adenine
ACP	Acyl carrier protein
ADP	Adenosine diphosphate
Akt	A protein family, which are also called protein kinases B (PKB) and play an important role in mammalian cellular signalling
AMP	Adenosine monophosphate (cAMP: Cyclic AMP: an important intracellular messenger in the action of hormones)
AMPK	5'adenosine monophosphate-activated protein kinase
ATP	Adenosine triphosphate: a high-energy compound that is the immediate source for muscular contraction and other energy requiring processes in the cell
ATPase	Adenosine triphosphatase: an enzyme that breaks down ATP to ADP and inorganic phosphate, releasing energy that can be used to fuel biological work
BCAA	Branched chain amino acid (includes leucine, isoleucine, and valine)
CaMK	Calmodulin kinase
CaN	Calcineurin
C	Cytosine
CCK	Cholecystokinin
CHO	Carbohydrate
CK	Creatine kinase: enzyme that catalyses the transfer of phosphate from phosphocreatine to ADP to form ATP
CNS	Central nervous system (brain and spinal cord)

C_aO_2	Content of oxygen in arterial blood
C_vO_2	Content of oxygen in venous blood (a bar over the v signifies mixed venous blood)
CO_2	Carbon dioxide
CoA	Coenzyme A that acts as a carrier for acyl groups (A stands for acetylation)
CoA-SH	Free form of coenzyme A
COOH	Carboxyl group
CoQ	Coenzyme Q or ubiquinone, an electron carrier that mediates transfer of electrons from flavoprotein to cytochrome c in the electron-transport chain
Cr	Creatine
dm	Dry matter
DNA	Deoxyribonucleic acid
DOPA	Dihydroxy phenylalanine
1,3-DPG	1,3-diphosphoglycerate
2,3-DPG	2,3-diphosphoglycerate
ETC	Electron-transport chain
FAD	Flavin adenine dinucleotide (oxidized form)
$FADH_2$	Flavin adenine dinucleotide (reduced form)
FDP	Fructose 1,6-diphosphate
FDPase	Fructose 1,6-diphosphatase
FFA	Free fatty acid
FMN	Flavin mononucleotide (oxidized form)
$FMNH_2$	Flavin mononucleotide (reduced form)
F6P	Fructose 6-phosphate
G	Guanine
GABA	Gamma-amino butyric acid
GDP	Guanosine diphosphate
GLUT-4	Glucose transporter found in cell membranes, including sarcolemma of muscle fibres
G1P	Glucose 1-phosphate
G6P	Glucose 6-phosphate
GTP	Guanosine triphosphate
H^+	Hydrogen ion or proton
HCO_3^-	Bicarbonate: the principal extracellular buffer
HDL	High-density lipoproteins
HK	Hexokinase: enzyme that catalyses the phosphorylation of glucose
Hz	Unit of frequency: cycles per second
5-HT	5-Hydroxy tryptamine or serotonin
IDL	Intermediate density lipoproteins
IGF	Insulin-like growth factor

IMP	Inosine monophosphate
J	Joule: unit of energy
kJ	Kilojoule: unit of energy ($kJ = 10^3\,J$)
K_m	Michaelis constant: the substrate concentration at which the velocity of an enzymatic reaction is half-maximal
LCAT	Lecithin–cholesterol acyl transferase
LDH	Lactate dehydrogenase: enzyme that catalyses the reversible reduction of pyruvate to lactate
LDL	Low-density lipoproteins
LPL	Lipoprotein lipase: enzyme that catalyses the breakdown of triacylglycerols in plasma lipoproteins
M	Molar: unit of concentration. (nM: nanomolar = 10^{-9}M; uM: micromolar = 10^{-6}M; mM: millimolar = 10^{-3}M)
MAPK	Mitogen-activated protein kinase
min	Minute: unit of time
mTOR	Mammalian target of rapamycin (mTOR), an intracellular signalling protein that integrates the input from multiple upstream pathways, including insulin and insulin-like growth factors. mTOR also functions as a sensor of cellular nutrient and energy levels and redox status
mole	Amount of substance represented by the molecular mass in grams
NAD$^+$	Nicotinamide adenine dinucleotide (oxidized form)
NADH	Nicotinamide adenine dinucleotide (reduced form)
NADP$^+$	Nicotinamide adenine dinucleotide phosphate (oxidized form)
NADPH	Nicotinamide adenine dinucleotide phosphate (reduced form)
NH$_2$	Amino group
NH$_3$	Ammonia
NH$_4^+$	Ammonium ion
O$_2$	Oxygen molecule
OH	Hydroxyl group
Pa	Pascal: unit of pressure. (One atmosphere = 101 kPa = 760 mmHg)
PCr	Phosphocreatine or creatine phosphate
PDH	Pyruvate dehydrogenase: the enzyme catalysing the conversion of pyruvate to acetyl CoA
PDK	Pyruvate dehydrogenase kinase
PEP	Phosphoenol pyruvate

PFK	Phosphofructokinase: the rate-limiting enzyme in glycolysis
pH	A measure of acidity/alkalinity. $pH = -\log_{10}[H^+]$
P$_i$	Inorganic phosphate
PCO$_2$	Partial pressure of carbon dioxide
PO$_2$	Partial pressure of oxygen
PPAR-γ	Peroxisome proliferator-activated receptor γ: one of a group of nuclear receptor proteins that act as transcription factors within cells, regulating the expression of genes
Q	Blood flow rate or cardiac output
R (group)	Organic side chain of an amino acid
RER	Respiratory exchange ratio
RNA	Ribonucleic acid (mRNA: messenger RNA; tRNA: transfer RNA)
ROS	Reactive oxygen species
RQ	Respiratory quotient
s	Second: a unit of time
SDH	Succinate dehydrogenase: an enzyme of the tricarboxylic acid cycle.
SOD	Superoxide dismutase
T	Thymine
TBARS	Thiobarbituric acid reactive species
TCA	Tricarboxylic acid
U	Uracil also international unit of enzyme activity: $1U = 1$ umol/min
UDP	Uridine diphosphate
UTP	Uridine triphosphate
VIP	Vasoactive intestinal peptide
VLDL	Very low-density lipoproteins
Vmax	Maximal velocity of an enzymatic reaction when substrate concentration is not limiting
VCO$_2$	Rate of carbon dioxide production
VO$_2$	Rate of oxygen uptake
VO$_2$max	Maximal oxygen uptake

Biochemical Terminology

Acid	A substance that tends to lose a proton (hydrogen ion)
Aerobic	Occurring in the presence of free oxygen
Allosteric enzyme	Enzyme that alters its three-dimensional conformation as a result of the binding of a smaller molecule (at a site different from its active site), often leading to inhibition or activation of its activity
Anaerobic	Occurring without the consumption of free oxygen
Anaplerotic reaction	A reaction that maintains the intracellular concentration of crucial intermediates that might otherwise become depleted (e.g. the formation of oxaloacetate from pyruvate by pyruvate carboxylase)
Base	A substance that tends to donate an electron pair or coordinate an electron
Buffer	A substance that in solution prevents rapid changes in hydrogen ion concentration (pH)
Calmodulin kinase	(CaMK) a family of different kinases can detect and respond to changes in calcium concentrations
Carboxylation	A reaction involving the addition of CO_2, catalysed by an enzyme using biotin as its prosthetic group.
Catecholamines	Biologically active hormones and neurotransmitters that are synthesized from the amino acid tyrosine. Examples include adrenaline, dopamine, and noradrenaline
cis-	A prefix indicating that geometrical isomer in which the two like groups are on the same side of a double bond with restricted rotation
Codon	The sequence of three base pairs in DNA or RNA that denotes a specific amino acid to be added during the synthesis of proteins
Coenzyme	Small molecules that are essential in stoichiometric amounts for the activity of some enzymes
Condensation	A reaction involving the union of two or more molecules with the elimination of a simpler group such as H_2O
Conformation	Shape of molecules determined by rotation about single bonds, especially in polypeptide chains about carbon–carbon links
Covalent bond	A chemical bond in which two or more atoms are held together by the interaction of their outer electrons

Covalent regulation	Control of enzyme activity by covalent bonding of phosphate groups to sites other than the active site of the enzyme
Cytokine	Cytokines are proteins or glycoproteins that act as signalling molecules allowing cellular communication
Deamination	Reaction involving the loss of an amino (NH_2) group
Decarboxylation	Reaction involving the loss of a CO_2 group
Dehydration	Reaction involving the loss of a water molecule
Eastern blotting	A technique to analyse proteins, lipids, or glycoconjugates, most commonly used to detect post-translational modifications to proteins
Enzyme	A protein with specific catalytic activity. They are designated by the suffix '-ase', frequently attached to the type of reaction catalysed. Virtually all metabolic reactions in the body are dependent on and controlled by enzymes
Epimerization	A type of asymmetric transformation in organic molecules
Flux	The rate of flow through a metabolic pathway
Free radical	Atoms, molecules, or ions that have one or more unpaired electrons that make these radicals highly chemically reactive
Genome	The genome is the entire DNA sequence of an organism and so essentially represents the genetic fingerprint of an individual. The term genome can be applied to all the variant forms of genes that exist in a population or species, and this can be termed the species genome
Geometrical isomerism	A form of sterioisomerism in which the difference arises because of hindered rotation about a double bond. An unsaturated fatty acid containing one carbon double bond has two isomers, depending on whether the hydrogen atoms are on the same (cis) or the opposite (trans) sides of the molecule
Gluconeogenesis	The synthesis of glucose from non-carbohydrate precursors such as glycerol, ketoacids, or amino acids
Glycogenolysis	The breakdown of glycogen into glucose-1-phosphate by the action of phosphorylase
Glycolysis	The sequence of reactions that converts glucose (or glycogen) to pyruvate

Glycosidic bond	A chemical bond in which the oxygen atom is the common link between a carbon of one sugar molecule and the carbon of another. Glycogen, the glucose polymer is a branched chain polysaccharide consisting of glucose molecules linked by glycosidic bonds
Half-life	Time in which half the quantity or concentration of a substance is eliminated or removed
Hydration	A reaction involving the incorporation of a molecule of water into a compound
Hydrogen bond	A weak intermolecular or intramolecular attraction resulting from the interaction of a hydrogen atom and an electronegative atom possessing a lone pair of electrons (e.g. oxygen or nitrogen). Hydrogen bonding is important in DNA and RNA and is responsible for much of the tertiary structure of proteins
Hydrolysis	A reaction in which an organic compound is split by interaction with water into simpler compounds
Hydroxylation	A reaction involving the addition of a hydroxyl (OH) group to a molecule
Ion	Any atom or molecule that has an electrical charge due to loss or gain of valency (outer shell) electrons
Ionic bond	A bond in which valence electrons are either lost or gained, and atoms which are oppositely charged are held together by electrostatic (coulombic) forces
Isoform	Chemically distinct forms of an enzyme with identical activities (also called isoenzyme), usually coded by different genes
Isomer	One of two or more substances that have an identical molecular composition and relative molecular mass, but different structure due to a different arrangement of atoms within the molecule
Isotope	One of a set of chemically identical species of atom that have the same atomic number but different mass numbers (e.g. 12-,13-, and 14-isotopes of carbon whose atomic number is 12)
Ketogenesis	The synthesis of ketones
Kinase	An enzyme that regulates a phosphorylation–dephosphorylation reaction
Lipolysis	The breakdown of triacylglycerols into fatty acids and glycerol

Metabolite	A product of a metabolic reaction
Mutation	Mutations occur when a gene is missing or defective, which means that the specific protein that it makes will also be missing or defective
Northern blotting	A technique used in molecular biology to detect RNA in a biological sample and thus gain information on gene expression
Nucleic acid	Nucleic acids carry the cell's genetic information in the form of long chains of nucleotides. The most common nucleic acids are *deoxyribonucleic acid* (DNA) and *ribonucleic acid* (RNA)
Nucleotides	Nucleotides are composed of a purine or pyrimidine base, a five-carbon sugar (either ribose (ribonucleotides) or 2′-deoxyribose (deoxyribonucleotides)), and between one and three phosphate groups
Oxidation	A reaction involving the loss of electrons from an atom. It is always accompanied by a reduction. For example pyruvate is reduced by NADH to form lactate. In the reverse reaction lactate is oxidized by NAD^+ when pyruvate is reformed
Peptide bond	The bond formed by the condensation of the amino group and the carboxyl group of a pair of amino acids. Peptides are constructed from a linear array of amino acids joined together by a series of peptide bonds
Phosphagen	The term given to both high-energy phosphate compounds, adenosine triphosphate and phosphocreatine
Phosphorylation	A reaction that involves the addition of a phosphate (PO_4^{2-}) group to a molecule. Many enzymes are activated by the covalent bonding of a phosphate group. The oxidative phosphorylation of ADP forms ATP
Prosthetic group	A coenzyme that is tightly bound to an enzyme
Rate-limiting enzyme	An enzyme in a metabolic pathway that regulates the slowest step in the pathway, and hence limits the rate of flux through the pathway
Reactive oxygen species	(ROS) ROS are highly reactive due to the presence of unpaired electrons. They are normally thought of as causing damage to cells by interacting with other molecules, but ROS they may have important roles in cell signalling

Reduction	A reaction in which a molecule gains electrons
Southern blotting	A method routinely used in molecular biology for detection of a specific DNA sequence in DNA samples
Stereoisomerism	The existence of different substances whose molecules possess an identical connectivity but different arrangements of their atoms in space
Substrate	The reactant molecule in a reaction catalysed by an enzyme
Thioester bond	A bond in which the oxygen has been replaced by sulphur. For example the linking of CoA in acetyl CoA is through a thioester bond
trans-	A prefix indicating a geometrical isomer in which like groups are on opposite sides of a double bond with restricted rotation
Transamination	Reaction involving the transfer of an amino (NH_2) group from an amino acid to a ketoacid
Transcription	The process by which RNA polymerase produces single-stranded RNA complementary to one strand of the DNA
Translation	The process by which ribosomes and tRNA decipher the genetic code in mRNA in order to synthesize a specific polypeptide or protein
Western Blotting	A molecular biology technique that allows separation of individual proteins and then used specific antibodies to detect and quantify specific protein molecules

Appendix 3
Units commonly used in biochemistry and physiology

Note that in these tables, the superscript '−1' convention has been used to indicate 'divided by' or 'per'. The alternative, as used in the text of this book, is to use a slash '/'. For example, to represent speed in metres (m) per second (s), either 'm s^{-1}' or 'm/s' can be used.

Table A3.1 Système International d'Unités (SI units).

Physical quantity	Name of unit	Symbol	Definition of unit
length	metre	m	
mass	kilogram	kg	
time	second	s	
temperature	kelvin	K	
	degree Celsius	°C	temperature K −273.15
amount of substance	mole	mol	
angle	radian	rad	
electric current	ampere	A	
potential difference	volt	V	$\text{kg m}^2\,\text{s}^{-3}\,\text{A}^{-1} = \text{J A}^{-1}\,\text{s}^{-1}$
electric charge	coulomb	C	A s
resistance	ohm	Ω	$\text{k gm}^2\,\text{s}^{-3}\,\text{A}^{-2} = \text{V A}^{-1}$
energy	joule	J	$\text{kg m}^2\,\text{s}^{-2}$
force	newton	N	$\text{kg m s}^{-2} = \text{J m}^{-1}$
power	watt	W	$\text{kg m}^2\,\text{s}^{-3} = \text{J s}^{-1}$
pressure	pascal	Pa	N m^{-2}
luminous intensity	candela	cd	
frequency	hertz	Hz	cycles s^{-1}
area	square metre	m^2	
volume	cubic metre	m^3	
density	kg per cubic metre	kg m^{-3}	
enzyme activity	katal	kat	mol s^{-1}

Physical quantity	Name of unit	Symbol	Definition of unit
length	angstrom	Å	10^{-10} m $= 10^{-1}$ nm
temperature	degree Fahrenheit	°F	1.8 temperature °C $+$ 32
energy	erg	erg	10^{-7} J
	calorie	cal	4.1868 J
	horsepower	hp	745.7 W
force	dyne	dyn	10^{-5} N
velocity		v	m s^{-1} and km h^{-1}
acceleration			m s^{-2}
pressure	bar	bar	10^5 N m^{-2}
	atmosphere	atm	101.325 kN m^{-2}
	torr	torr	133.322 N m^{-2}
volume	litre	l	10^{-3} m^3
density			g cm^{-3} = g ml^{-1}
enzyme activity	international unit	IU or U	μmol min^{-1}
concentration	[ ]	mol l^{-1} or M	mol l^{-1}
viscosity	poise	P	10^{-1} kg^{-1} s^{-1} = 10^{-1} Pa s
radioactivity	curie	Ci	37×10^9 counts s^{-1}
	roentgen	R	22.58×10^{-4} count s kg^{-1}
	rad	rad	0.01 J kg^{-1}

Table A3.2 Derived SI units and non-SI units allowed or likely to be met.

	Prefix	Symbol	Example
Fraction			
10^{-1}	deci	d	decilitre, dl
10^{-2}	centi	c	centimetre, cm
10^{-3}	milli	m	millisecond, ms
10^{-6}	micro	μ	micromole, μmol
10^{-9}	nano	n	nanometre, nm
10^{-12}	pico	p	picogram, pg
10^{-15}	femto	f	femtolitre, fl
Multiple			
10^3	kilo	k	kilogram, kg
10^6	mega	M	megajoule, MJ
10^9	giga	G	gigaohm, GΩ

Table A3.3 SI fractions and multiples.

Index